Territory: the Claiming of Space

David Storey

Geography Department
University College Worcester

Prentice
Hall

An imprint of Pearson Education

Harlow, England · London · New York · Reading, Massachusetts · San Francisco
Toronto · Don Mills, Ontario · Sydney · Tokyo · Singapore · Hong Kong · Seoul
Taipei · Cape Town · Madrid · Mexico City · Amsterdam · Munich · Paris · Milan

Pearson Education Limited
Edinburgh Gate
Harlow
Essex CM20 2JE
England

and Associated Companies throughout the world

Visit us on the World Wide Web at:
http://www.pearsoneduc.com

First published 2001

© Pearson Education Limited 2001

The right of David Storey to be identified as author of
this Work has been asserted by him in accordance with
the Copyright, Designs and Patents Act 1988.

ISBN 0 582 32790 3

British Library Cataloguing-in-Publication Data
A catalogue record for this book is available from the British Library

Library of Congress Cataloging-in-Publication Data
Storey, David, 1960–
    Territory : the claiming of space / David Storey.
        p.   cm. — (Insights in human geography)
    Includes bibliographical references and index.
    ISBN 0–582–32790–3 (pbk.)
    1. Political geography.   2. Nationalism.   3. National state.   4. Territory, National.
    I. Title.   II. Insights into human geography.
    JC319.S76   2000
    320.1'2—dc21   00–046949

10 9 8 7 6 5 4 3 2 1
05 04 03 02 01

Typeset by 35 in 11/12pt Adobe Garamond
Produced by Pearson Education Asia Pte. Ltd.
Printed in Singapore

Territory

This volume is part of a series of new human geography teaching texts. The series, *Insights in Human Geography*, is designed as an introduction to key themes in contemporary human geography. Together, the volumes form the basis for a comprehensive approach to studying human geography. Each volume, however, is free-standing and can be studied on its own as an introduction to a specific sub-field within the discipline.

The series is built around an exploration of both the substantive geographies of the real world and the conceptual and theoretical frameworks that are required to contextualize them. Each volume not only provides a thorough grounding in one particular sub-field of human geography but also looks to the broader picture, providing students with a geographical perspective on contemporary issues, and showing how recent changes in the real world have led to changes in the ways that geographers approach and understand the world.

Contents

List of plates

List of figures

List of tables

Author's acknowledgements

The writing of this book has taken much longer than originally envisaged and I am extremely grateful to those who assisted in speeding it towards completion. Particular thanks are due to Heather Barrett, John Dixon, Ronnie Kowalski, Paddy McNally and Richard Yarwood who read drafts of various chapters and who offered helpful advice. I am especially grateful to Andy Storey and Anne Sinnott who read the entire manuscript and commented critically on it. The series editors Paul Knox and Susan Smith offered helpful advice. I am grateful to Jane Coulter for preparing the index. Thanks are due to Alan Bennett for his assistance with many of the maps and to John Dixon for Plate 7.1, Liam O'Hare for Plates 8.3 and 8.4, Jennifer Storey and Fergus Sullivan for Plate 8.5. Matthew Smith and Tina Cadle at Pearson Education Limited provided constant encouragement and gentle reminders. Thanks to David Norris and Caroline Ellerby for copy-editing and proof-reading. I am grateful to my colleagues at University College Worcester for their friendship, advice and support. Finally, thanks to Anne for encouraging, supporting and listening.

David Storey
September 2000

Publisher's acknowledgements

We are grateful to the following for permission to reproduce copyright material:

Figure 1.2, Figure 3.4 and Figure 6.1 from *Political Geography. World-Economy, Nation-State and Locality*, Fourth Edition, Prentice Hall (Taylor, P.J. and Flint, C. 2000); Figure 3.3, Figure 5.1, Figure 7.1 and Figure 7.2 from *Federalism. The Multiethnic Challenge*, Longman (Smith, G. ed.); Figure 8.1 from *Urban Social Geography. An Introduction*, Third Edition, (Knox, P. 1995) and Figure 8.2 from *The American Urban System. A Geographical Perspective*, Longman (Johnston, R.J. 1982) all reproduced with permission from Pearson Education Ltd; Figure 1.3 from *An Introduction to Political Geography*, Second Edition, Routledge (Short, J.R. 1993); Figure 5.3 from *In Search of Ireland. A Cultural Geography*, Routledge (Graham, B. (ed.) and Figure 7.5 from *Town and Country Planning in the UK*, Twelfth Edition, Routledge reproduced with permission of Taylor & Francis Books Ltd (Cullingworth, J.B. and Nadin, V. 1997); Figure 2.3 from *The Colonial Empires. A Comparative Survey from the Eighteenth Century* reproduced with permission from Macmillan Ltd (Fieldhouse, D.K. 1982); Figure 3.2, Figure 3.5, Figure 3.6 and Figure 7.7 from *Political Geography* copyright © 1993, reprinted by permission of John Wiley & Sons, Inc. (Glassner, M.I. 1993); Figure 3.8 'The political landscape of partition. The case of Cyprus' *Political Geography* 16 (6) from (Kliot, N. and Mansfield, Y. 1997) and Figure 7.8 from 'Voluntary action in rural areas: the case of Neighbourhood Watch' *Journal of Rural Studies* 11 (4) reprinted with permission from Elsevier Science (Yarwood, R. and Edwards, B. 1995); Figure 4.2 reproduced with permission from the British National Party; Figure 4.3 from *A Geography of the Welsh Language, 1961–1991* reproduced with permission from University of Wales Press (Aitchison, J. and Carter, H. 1994); Figure 5.2 from *The Fall of Yugoslavia. The Third Balkan War*, Third Edition, Penguin Books 1992, Third revised edition 1996 copyright © Misha Glenny, 1992, 1993, 1996, reproduced by permission of Penguin Books Ltd (Glenny, M. 1996); Figures 6.2 and 6.3 from *Global Shift. Transforming the World Economy*, Third Edition reprinted by permission of Paul Chapman Publishing and the author (Dicken, P. 1998); Figure 6.6 based *The United States and Africa. A History* reprinted with permission from Cambridge University Press on (Duignan, P. and Gann, L. 1985); Figure 8.3 reproduced by permission of the Ordnance Survey of Northern Ireland on behalf of the Controller of Her

Majesty's Stationery Office © Crown copyright permit number 1592; Figure 8.4 from *Gendered Spaces* copyright © 1992 by the University of North Carolina Press, used by permission of the publisher (Spain, D. 1992); Figure 8.6 from 'West Hollywood as symbol: the significance of place in the construction of gay identity' *Environment and Planning D: Society and Space* 13 (2) © Pion, London, reproduced with permission from (Pion Ltd Forest, B. 1995); Table 8.1 from Boal, F.W. (1978) 'Ethnic residential segregation' in Herbert, D.T. and Johnston, R.J. (eds) *Social Areas in Cities* © John Wiley & Sons Ltd, reproduced with permission; John Fiddy Music for permission to reproduce lyrics from the Austrian national anthem; Plate 5.2 *Hereford Dynedor and the Malvern Hills from Haywood Lodge*, painting by George Robert Lewis, 1815, reprinted by permission from Tate Gallery, London 2000; Plate 5.3 Ironbridge Gorge postcard, copyright J. Salmon Ltd., Sevenoaks, England ©; Plate 6.2 cardboard cut-out of AK-47 attached to lampost in a republican area, Newry, Northern Ireland, photo by Paul Anthony McErlane; Plate 8.1 from Ingrid Pollard.

Whilst every effort has been made to trace the owners of copyright material, in a few cases this has proved impossible and we take this opportunity to offer our apologies to any copyright holders whose rights we may have unwittingly infringed.

Chapter 1

Introduction

Politics and political relationships underpin the world in which we live. From a geographical perspective, the most obvious manifestation of this is the division of the earth's surface into political–territorial units known as states. The term 'territory' is most usually used in reference to the area of land claimed by a country. However, territories exist at a variety of spatial scales from the global down to the local. Territory refers to a portion of geographic space which is claimed or occupied by a person or group of persons or by an institution. It is, thus, an area of 'bounded space'. The process whereby individuals or groups lay claim to such territory is referred to as 'territoriality'. The investigation of various dimensions of territorial formation and behaviour at different spatial scales forms the central focus of this book. It examines the construction of territories and the use of territorial strategies at the level of the state, substate divisions and at the micro-scale level in individual urban areas, in the workplace and within the home. The next two sections of this chapter provide a brief introduction to these themes. The purpose of the book is to explore how territoriality is used as a strategy to assert power or to resist the power of a dominant group and the penultimate section of the present chapter expands on this idea. The chapter concludes by providing an outline of the book's structure.

Territory at a macro-scale

At a global level the pursuit, by major powers, of spheres of influence, whether in the eras of formal colonialism, superpower tension or other forms of geopolitical rivalry, represents a distinct version of territorial behaviour. In Europe, following the demise of feudalism, older political entities were replaced initially by a series of city-states and larger territorially based units. In fact, the origins of the word 'territory' can be traced back to medieval times. In the Roman era the word *territorium* was associated with both community and territory. Slowly, the idea of owing allegiance to the territory began to supersede allegiance to lord or to God. Wars began to be fought in the name of territorial formations. This is not to suggest some significant difference in the underlying purpose. Wars are fought over issues of power and control but the significance arose from the fact that now they were being fought utilizing territorial terminology. Territory had become the mobilizing force and, to a

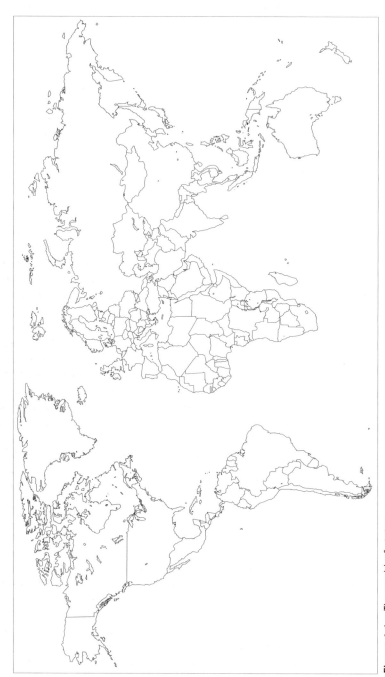

Figure 1.1 The world of states.

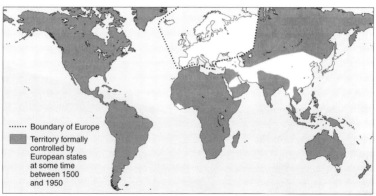

Figure 1.2 The extent of European territorial control.
Source: Taylor and Flint, 2000

considerable extent, control of territory became the geographical expression of political power.

Ultimately, the present interstate system evolved into a more cohesive territorial political system in which all the world's territory is seen to belong in distinct political entities commonly known as states (Figure 1.1). This process was facilitated by the advent of imperialism, a political project whereby a number of European countries began a process of territorial accretion beyond the confines of Europe through the process of acquiring colonies. Empire-building by European powers, most notably Spain, Portugal, The Netherlands, France and England, represented a territorial expression of power on the part of these countries which led to much of the world's surface area being carved up between them (Figure 1.2). This has left an enduring territorial legacy (together with a range of economic, political and cultural consequences) throughout Africa, Asia and the Americas.

More recently, the Cold War era of the middle years of the twentieth century saw the United States and the Soviet Union establishing, or seeking to establish, geographical spheres of influence. This was most notable in the Soviet Union's effective control of eastern Europe and the USA's flexing of its political and military muscle in its own 'backyard' of central America. While this might not have been direct imperialism and ultimate formalized territorial control, it was a process whereby 'friendly' governments were encouraged, or sometimes forcibly installed, in countries which the superpowers saw as vital to their strategic interests. In this way, the world could be seen as divided between those who were allied to one side or the other, with a small group of neutral countries (Figure 1.3).

The evolution of trading blocs such as the European Union (EU) and the Association of South East Asian Nations (ASEAN) represents yet another form of territorial organization at a global scale. The current trajectory of the EU, from a trading bloc, through an economic union and towards greater and greater political integration, has been allied to a policy of continued spatial expansion. To some, this type of suprastate confederation is seen as the

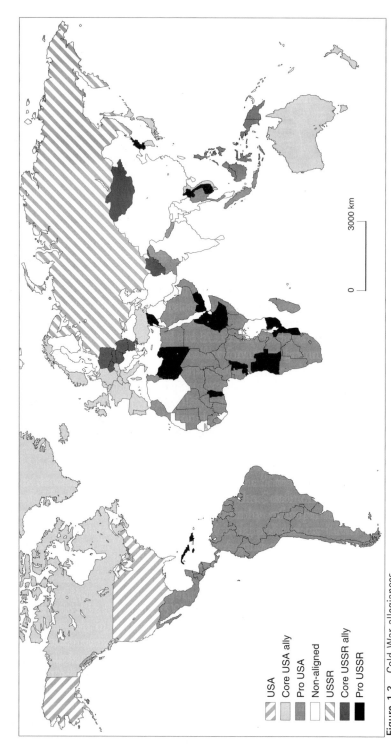

Figure 1.3 Cold War allegiances.
Source: Short, 1993

successor to the state system in an increasingly globalized world where state boundaries are assumed to have diminishing significance.

Below this global level in the spatial hierarchy there exists the state system already referred to; the most obvious example of formalized territorial organisation in the world today. The political division of the planet into territorial states is a well-known and 'taken for granted' feature. There is a general acceptance of the idea of formalized and rigidly demarcated boundary lines even though the precise location of these may be subject to dispute in certain instances. This process of state formation has been facilitated through the concept of the nation; a territorial ideology linking people to place.

Territoriality can also be observed at the substate level. The internal divisions of a country, whether through local or federal government, represent a formalized internal territorialization. German *länder*, English counties or the individual states of the USA are all substate territorial formations which are integral parts of the larger polity. Recent trends towards 'rolling back the state', together with an increased emphasis on community-based responses to social and economic concerns, have led to a growing importance being attached to quasi-governmental institutions and to locality-based organisations. These 'community' responses to certain issues represent another version of territorialised political behaviour, whereby spatially defined communities seek solutions to problems within their own localities.

Territory at a micro-scale

The instances mentioned so far are the more obvious examples of territorial formations. However, Cox (1991) has pointed to the increasing utilization of concepts of territory and territoriality by geographers concerned with applying social theory within the discipline. This has led to a consideration of territorial behaviour and strategies in contexts not directly connected to the state. In other words, there has been a broadening out from what might be seen as the traditional base of political geography. This leads to a consideration of territory as any (even informally) bounded space and a recognition that such spaces may experience internal or external contestation.

Within urban areas, clear spatial divisions can be seen to exist with fault lines dividing zones in terms of affluence, class or ethnicity. As a result, particular areas become characterized in certain discourses as, for example, 'poor', 'working class' or 'Asian'. Such division of urban space into informal territories leads to a consideration of the ways in which space can be seen as gendered or racialized. For example, ethnicized ghetto formation, a clear territorial feature, occurs in many urban areas. These ghettos often come to be seen as the territory of the particular ethnic group concentrated there. Similarly, other spaces may be divided in terms of gender. For example, football pitches (and indeed football grounds generally, not just the pitch) have traditionally been seen as male spaces while the kitchen has been seen as a female space.

Owning a house can also be seen in territorial terms. Private property is regarded by many as a logical and necessary outcome of human territorial behaviour. It represents a claim to space which is reinforced by the legal system of many countries. Taking the idea of territory down to its most elementary level, the desire for personal space can be seen as a form of territorial behaviour. Humans like to have a pocket of space around them which is 'theirs' and they resent others 'invading' their space. This can be interpreted as a territorial claim to a portion of geographic space.

Territory and power

Contemporary concerns within geography extend well beyond mere descriptions of territorial formation and behaviour. There is a need to explain why such behaviour occurs. It is the central premise of this book that territorial strategies are employed by individuals or groups in order to attain or to maintain control. Whether explicit or implicit, control over territory is a key political motivating force and the apportioning of space or specified territory results from the interplay of social and political forces. Territoriality can be seen as the spatial expression of power. The processes of control and the contestation over particular territory are thus a key element in what is known as political geography. Any consideration of territories necessarily raises questions to do with boundaries. The borders of a territory may be quite clearly defined and formalized, as between most countries, or they may be somewhat informal or ill defined, as is the case when considering the territories of rival urban gangs whose 'patch' or 'turf' may well have a core but the precise boundaries of which may be unclear. Territoriality and the imposition of boundaries are political strategies designed to attain particular ends. From all of this it is obvious that territories are often the centre of disputes (in many instances quite violent disputes) and the areal extent of particular territories (and even their right to exist) is quite often contested.

Recent writing in cultural geography uses ideas of social and spatial exclusion and notions of transgression in highlighting unequal access to particular spaces. There is a growing interest in the ways in which 'others', or non-dominant groups, are seen to be excluded while these groups, in turn, resist those processes which actively exclude them. As will be seen in the examples which follow, non-dominant groups often attempt to wrest control of geographical space in order to assert their presence. People resist power and the imposed territorial boundaries through which that power is expressed. The borders erected around territories, whether formal or informal, are subject to periodic transgression. In practical terms it may make a considerable difference which side of the line a person finds themselves on: hence the desire by many Mexicans to cross the Rio Grande into the United States or the risks taken by those who crossed, or attempted to cross, the Berlin Wall. Such strategies are set against the background of a 'postmodern' world in which, some argue, boundaries are becoming more and more permeable and ephemeral as processes of globalization increase in intensity and as formal state borders are seen as being of less and less importance. As a consequence, territories are not static and are constantly produced and re-produced. While efforts are constantly made to reinforce territorial control, these are often resisted and space is continually transgressed.

While highlighting the functions of territory, the phenomenon should not be reduced to a mere commodity or be seen simply as a spatial container. As will become clear territory can play an important role in the formation of peoples' self-identity. The chapters which follow focus on the extent to which territorial identities are important components in our overall sense of identity. The book explores the theoretical underpinnings lying behind different expressions of territoriality. This theoretical material is supported by a variety of historical and contemporary examples ranging from consideration of overt political conflicts, such as those surrounding the north of Ireland and Israel–Palestine, to issues such as local government and domestic space.

While it is obvious that the processes giving rise to national political conflicts are very different from those associated with divisions of space in the domestic sphere, nevertheless they both pertain to human attempts to control portions of space or to resist that control. As such, the many diverse forms of territorial behaviour considered in this book are connected in the sense of demonstrating peoples' interaction with place and space. It should also be obvious that the division between macro-scale and micro-scale is somewhat artificial. Many macro-level processes, such as state-building, have quite localized impacts while events in particular places may well impact at more global levels. The macro and the micro are not structurally separate geographic spaces. In summary there is a concern with the processes through which territories are produced, the contested nature of territorial formation and control and the importance of territorial identity. These form the key themes which underpin the book. As a result, the ideas of power, dominance, resistance, conflict and identity run throughout each chapter.

Structure of the book

Chapter 2 explores the concept of territory and provides an overview of theories explaining the existence of human territoriality. The direct political control over a designated territory and the process of state formation is the subject of Chapter 3. Associated with state formation is the territorial ideology of nationalism. Concepts of nation and nationalism, and the symbolism associated with them, form the subject matter of Chapters 4 and 5 in the book. Chapter 6 deals with the debates surrounding the future of the state as a sovereign territorial entity. Territoriality at the substate level, as evidenced through internal subdivisions, is dealt with in Chapter 7. The more micro-scale versions of territoriality – those which occur at the level of individual urban areas, for example the 'racialization' of space by particular ethnic groups (whether dominant or subordinate) or the informal claiming of space on the basis of gender – are explored in Chapter 8. The chapter also deals briefly with territorial behaviour at the level of the home or the workplace.

In focusing on formal and informal territorial strategies, and in emphasizing the contested nature of territory, the book deals with aspects of what might be seen as the traditional subject matter of political geography, namely a concern with the state and the nation. However, in exploring other dimensions of territorial control and focusing on various spatial strategies of dominance and resistance, the book incorporates recent developments in human geography and places these within a consideration of power relationships considered from a spatial perspective.

Further reading

Suggested reading on the issues dealt with in this book is provided at the end of each chapter. For overviews of the evolution of a territorial–political system, see P.J. Taylor and C. Flint, *Political Geography. World-Economy, Nation-State and Locality* (4th edition, Prentice Hall, Harlow, 2000). On geopolitics, see the work of John Agnew, including *Geopolitics. Revisioning World Politics* (Routledge, London, 1998) and (with Stuart Corbridge) *Mastering Space. Hegemony, Territory and International Political Economy* (Routledge, London, 1995). See also the various writings of Gearoid Ó Tuathail, including *Critical Geopolitics. The Politics of Writing Global Space* (Routledge, London, 1996) and the edited volume (with Simon Dalby and Paul Routledge) *The Geopolitics Reader* (Routledge, London, 1998). See also Klaus Dodds, *Geopolitics in a Changing World* (Prentice Hall, Harlow, 2000).

General introductory geography texts which can usefully be consulted alongside this volume include P. Cloke, M. Crang and M. Goodwin (eds), *Introducing Human Geographies* (Edward Arnold, London, 1999), P. Cloke, C. Philo and D. Sadler, *Approaching Human Geography* (Paul Chapman, London, 1991), R.J. Johnston, *Geography and Geographers. Anglo-American Human Geography since 1945* (4th edition, Edward Arnold, London, 1991), D.N. Livingstone, *The Geographical Tradition. Episodes in the History of a Contested Enterprise* (Blackwell, Oxford, 1992), R. Peet, *Modern Geographical Thought* (Blackwell, Oxford, 1998) and J. Allen and D. Massey (eds), *Geographical Worlds* (Oxford University Press/Open University, Oxford, 1995).

Chapter 2

Territory and territoriality

As indicated in Chapter 1, there is an observable tendency for humans to engage, either individually or collectively, in forms of territorial behaviour. We claim space and we interact with space and place on an everyday basis. This chapter introduces ways of conceptualizing the issue of territoriality in human societies. Broadly, current thinking on territoriality reduces to two main sets of theories. Firstly, there are biological and genetic theories which argue that territoriality is an innate feature of all species, including humans, and that forms of territorial behaviour are, therefore, natural. Juxtaposed to these theories is a body of thought which rejects this naturalizing of territoriality and suggests that territorial behaviour in humans is a phenomenon arising out of our broader socio-political conditioning. Each of these broad sets of theories is outlined in the sections which follow. It is important to bear in mind that these theoretical overviews are presented in somewhat simplistic ways in order to convey their essence. It should also be borne in mind that there are a variety of theoretical positions which exist somewhere on a spectrum ranging from totally biological theories through to totally 'conditioned' theories. Many would argue that an understanding of territorial behaviour requires a theory situated somewhere between these polar positions (Figure 2.1).

Figure 2.1 Theories of territoriality.

Biological and genetic approaches

Traditionally, the discussion of territoriality has been led by biologists, anthropologists and psychologists. Given their biological basis, it is not surprising that many of these arguments tend to support the idea of territorial behaviour as 'natural' rather than learned or contrived. Essentially two key positions emerge out of this. Firstly, there is a deterministic perspective which sees the acquisition of territory as a natural phenomenon. The second more nuanced perspective sees territorial behaviour in a behaviourist but non-deterministic sense.

In its crudest form, the deterministic argument holds that the need for space is a characteristic innate to all species, including humans. When allied to theories suggesting that aggression is a natural phenomenon (Lorenz 1966), this desire for space leads 'naturally' to the acquisition of territory, by the use of aggressive behaviour if necessary. There is, then, an impulse to defend this territory against others seeking to 'invade' it. Utilizing ethology (the study of animals in their natural environment), a set of arguments has been developed whereby human behaviour is seen to mirror that of animals. These bio-ethological views have been widely disseminated through the writings of people such as the anthropologist Robert Ardrey who argues that animals and humans behave in an intrinsically territorial manner. Humans have what Ardrey (1967) describes as a 'territorial imperative' which compels them to defend space. This is a crude biologically determinist position from which territorial behaviour in humans is seen as a natural and unchanging phenomenon (see Box 2.1).

Parallels between the behaviour of humans and animals have been popularized by the zoologist Desmond Morris in his many books, such as *Manwatching* (1973) and *The Naked Ape* (1994). Morris argues that humans are simply another species of animal and, as such, human behaviour patterns are primarily a result of genetic programming. One of these behavioural traits is the defence of territory. Humans are, in the view of Morris, territorial creatures. Relying heavily on the work of Lorenz, Morris argues that we have an innate need to defend territory, whether at the level of the nation or at the micro-scale level of our daily habitat, i.e. the home. Morris sees our tendency to adorn our homes in particular ways as a means of asserting our individuality through placing our 'stamp' on our own territory. In this way the simple act of painting the front door can be read as an assertion of 'territorial uniqueness' (Morris 1994: 124).

The socio-biologist Richard Dawkins, in his popular book *The Selfish Gene* (1976), argued that humans act as mere containers for our genes. Simplistic interpretations of this argument suggest that people are genetically programmed to defend those who are like themselves. This leads to territorial defence against people who are different. In this way violent conflict and phenomena such as selfishness and xenophobia are seen as natural and inevitable.

These analyses of territorial behaviour, it has been argued, hold true both at the level of the individual defending her or his property and at the collective level of the state defending itself against its rivals. One geographer who leaned heavily towards biological theories was Friedrich Ratzel. In Germany, in the second half of the nineteenth century, he developed organic theories of state formation. These will be returned to in Chapter 3. Suffice to say for the moment that Ratzel's thinking led him to view state expansion as a necessary means of state survival. He borrowed from Darwin's evolutionary theory in suggesting that states might need to adopt a 'survival of the fittest' strategy in order to retain power. This serves to justify aggressive strategies of territorial

Box 2.1 Territoriality and biological determinism

Biological determinism rests on the idea that humans' genetic make-up is the key influence on our behaviour. While environmental determinism is a paradigm which suggests that human behaviour is moulded by the physical environment in which we live, biological or genetic determinism suggests that our behaviour is governed by our DNA. The argument is that humans are genetically programmed to behave in particular ways. Many of the arguments used derive from ethology. This is the study of the behaviour of animals in their natural environment (not in a laboratory).

The anthropologist Robert Ardrey published a book in 1967 entitled *The Territorial Imperative*. Its subtitle *A Personal Inquiry into the Animal Origins of Property and Nations* provides a clear guide to its author's understanding of the causes of territorial behaviour. Ardrey argues that animals and humans have an in-built territorial urge and have, therefore, an inner compulsion to possess and to defend space. Using a variety of selectively chosen examples from the animal world, Ardrey argues that our behaviour is controlled by our genes and that we are ultimately unable to help ourselves. In essence, Ardrey's book boils down to an ideological critique of communism (on the basis that collective endeavours are against nature) and a defence of capitalism and private property (on the basis that humans are naturally individualistic rather than co-operative).

Desmond Morris, a zoologist, has written a large number of books and presented television programmes dealing with aspects of human behaviour. As with Ardrey, Morris' bio-ethological stance results in a very uncritical linking of human and animal behaviour. In *The Soccer Tribe*, a study of football published in 1981, he suggests that the behaviour of players and fans alike mirrors entirely behaviour from the animal world. From such studies, people like Ardrey and Morris have asserted that human behaviour is essentially pre-programmed. These determinist arguments tend to set limits to human behaviour rather than see human existence as one of possibilities. They do not consider the many ways in which our broader social, economic, political and cultural environments might shape our development and, hence, our behaviour.

defence and acquisition. More recently, the geographer Malmberg, from a biologically determinist position, asserted that 'territorial defence is based on this instinctive aggression' (1980: 26).

A more nuanced argument suggesting an innate basis for territorial behaviour derives from the work of the psychologist Piaget who examined children's need for security. Children find security in that which is familiar (Piaget and Inhelder 1967). Thus, they need space in which they feel comfortable and safe and which they can regard as their own. Assuming this requirement stays with us as adults, then it follows that there is a need to claim our space and to assert some form of control over it, either directly or indirectly.

The arguments outlined above form a component in the wider debate between innate and learned behaviour or between nature and nurture. Determinist arguments follow Freud in attributing many human characteristics to nature. From this perspective, territorial behaviour is viewed as being innate to humans rather than being a product of our social and cultural conditioning. These arguments have a certain appeal in that they appear to provide an explanation for elements of observable human behaviour, whether they be international conflicts or the activities of rival football fans. In essence, the argument is that territorial behaviour is a consequence of our evolutionary past rather than our cultural present.

While not rejecting totally the notion of parallels between the behaviour of humans and other species, there are a number of criticisms which can be made. Firstly, the arguments used by Ardrey and others could be seen as bad science. There are three main points here. The first is that, in many instances, there is a distinct lack of empirical evidence to support the suggestions being made. This renders the conclusions more akin to assertions than to meaningful interpretations. The second point is that the examples used are highly selective. Ardrey uses examples which support his case. He might have used others which would cast grave doubts on his assertions. The third 'scientific' weakness is the huge leaps in logic employed by Ardrey and others. Most biologically determinist arguments are predicated on the assumption that human behaviour can be extrapolated from the behaviour of certain animals. This is more a leap of faith than a scientifically proven fact. It has been argued that such extrapolation is questionable owing to differences in cognitive abilities between people and animals, humans' capacity for language and culture, and the sheer diversity of animal species and human behaviour (Hinde 1987). The extent to which it is valid to infer group behaviour on the basis of a small number of individual occurrences is also open to question. In short, Ardrey tends to assert rather than prove the existence of a territorial imperative. In doing so, he appears to dispense with the academic objectivity which he otherwise appears to laud.

A second type of criticism relates to apparent contradictions in some of the arguments, or at least to the fact that the evidence may be open to alternative interpretations. To amass evidence of territorial behaviour in humans is not in itself proof that we are innately territorial. It might just as easily be seen as proof that we are all conditioned in broadly similar ways. In rejecting overtly biologically determinist theories, the anthropologist Alland argues that 'human nature is largely open' (Alland 1972: 24). He suggests that there is a false dichotomy between biological and social conditioning; rather, he argues, the two interact. As he sees it 'in part, man [sic] adapts biologically to his environment in a non-biological way – through culture' (Alland 1972: 22). He appears to suggest a form of possibilism whereby humans, rather than being programmed to respond in particular ways, can decide on a course of action dependent on the opportunities available to them. Similarly, the evolutionary biologist Stephen Jay Gould (1983, 1991) rejects the binary opposition between nature and nurture and argues that, while biology is hugely influential, it presents a series of opportunities rather than setting limits to our behaviour.

The fact that different cultures have different modes of behaviour leads to the suggestion that the supposed innate basis of human territoriality is less strong than Ardrey, Morris and others would have it. Morris (1973) talks of the human need for personal space but in doing so he points to the fact that the amount of space needed varies from society to society, a fact noted long ago by Hall (1959). This suggests that social or cultural conditioning (the nurture side of the nature–nurture debate) plays a role in explaining territorial behaviour.

A third strand of criticism relates to the obvious ideological underpinnings of the arguments used by some, most notably Ardrey. Biologically based arguments often (wittingly or unwittingly) lend support to particular ideological positions (Rose, Kamin and Lewontin 1990). The arguments of Ardrey and Morris, for example, tend to justify aggression and 'naturalize' conflict. Animosity towards particular ethnic groupings or towards others recognizably 'different' may then be excused, thereby justifying racist or sexist behaviour. In addition, in presenting territorial defence as a natural phenomenon, they provide useful support for the institution of private property. Humans' desire to own and possess is seen as natural rather than a cultural product. Thus, Ardrey criticizes collective farms in the then Soviet Union on the grounds that their inefficiency derives from the fact that peoples' natural connection or affiliation with territory has been broken. In this way, private property becomes naturalized; it is seen as a consequence of our genes rather than as a function of political relationships. This can be interpreted as a biological justification of an ideological position. Slowe (1990) traces this debate (in so far as it impinges on territory and property rights) back to Locke and Rousseau. The former saw private property as natural while the latter argued that it was in fact against nature. In this instance, the 'natural' arguments are used to support a culture of individuality and competition, as opposed to one of mutuality and co-operation. Gould has referred to biological determinism as a 'sociopolitical doctrine masquerading as science' (1991: 302).

From the above criticisms it follows that more sensitive theories are required. A number of social psychological and ethological theories have been advanced which suggest that human and animal behaviour might be seen as analogous rather than homologous, as Ardrey and others suggest. In other words, there may well be behavioural parallels between animals and humans but this is not proof that the motivating factors are the same. Many social and environmental psychologists would argue that behavioural similarities do not necessarily reflect similar processes. Gold suggests that

> territoriality represents a culturally derived and transmitted answer to particular human problems, not the blind operation of instinct . . . Its rules, mechanisms, and symbols are developed gradually over time and are passed from one generation to the next by the . . . process of socialization. (Gold 1982: 48)

Given the criticisms outlined above, it is obvious that alternative explanations of territorial behaviour need to be explored. There is a need to move beyond explanations based largely on genetics to ones which consider the social context in which humans exist.

Socio-political theories _____

There is a major strand of thinking in social science which argues that human behaviour, far from being innate, is heavily conditioned by the wider economic, social, political and cultural environment in which we exist. It is suggested that much of human behaviour is learned and not natural. These arguments lead to a consideration of socio-political influences. With respect to territoriality, a body of thought exists which rejects the determinism of many psychological and ethological arguments and which suggests that human territorial behaviour is a product of conditioning rather than a biological urge. Central to this is an emphasis on power relationships rather than biology.

Two geographers who have contributed to this debate are Jean Gottman and Robert Sack. Gottman asserted that territoriality refers to 'a relationship between a community of politically organised people and their space'. He argued that 'civilised people [*sic*] . . . have always partitioned the space around them carefully to set themselves apart from their neighbours' (Gottman 1973: 1). He highlighted the fact that territory, as he saw it, represented a portion of geographical space under the jurisdiction of certain people. Its wider significance thus arises from the fact that 'it signifies also a distinction, indeed a separation, from adjacent territories that are under different jurisdictions' (Gottman 1973: 5). Thus, territory, and the assertion of control over it, represents an expression of power: 'we control this space'.

Gottman identified two reasons for territoriality. Firstly, it confers security. Territory can be converted into defensible space. Secondly, he argued that it can provide opportunities through allowing the economic organization of space. This allows people to pursue 'the good life'. In this, he was referring back to both Plato and Aristotle. Plato was concerned with the creation of territorial units which would be self-sufficient and secure. There was also a moral dimension to Plato's thought. He felt that moral goodness could be attained–maintained through this territorial system, idealized in the city-states or *polis* of medieval Europe. Aristotle, on the other hand, argued that territorial units should not isolate themselves but should actively participate in trade. In this way, the formalizing of a territorial system offered opportunities for advancement and not just security. This consideration of the issues of security and opportunity suggests a political and economic basis for territorial formation. Territorial states allow for a standardization of currency and economic regulations and may be seen as a more efficient means of political organization than systems of overlapping jurisdictions.

A much more overt political reading of territoriality is provided by the geographer Robert Sack. He views territoriality and the defining of space as a reflection of power (see Box 2.2). Territoriality is defined by Sack as 'the attempts by an individual or group to affect, influence or control people, phenomena, and relationships, by delimiting and asserting control over a geographic area' (Sack 1986: 19). Control of space can be used in order to affect, influence or control resources. Power is exerted over individuals whether through controlling the behaviour of those in a specified territory or through excluding people

---

Box 2.2   Robert Sack

Surprisingly, given geography's pre-occupation with space and place, Robert Sack is one of the few geographers to have attempted a serious analysis of the concept of human territoriality. His book *Human Territoriality. Its Theory and History* was published in 1986 and follows on an earlier paper (Sack 1983). In it, Sack rejects determinist theories of territoriality in favour of a 'political' theory which sees territorial behaviour as a geographic strategy rather than a basic instinct. Space and society are interconnected and territoriality is the process which connects them.

Sack sees territories as social constructs rather than natural phenomena. Using both historical and contemporary examples, he argues that territoriality is embedded in social relations. He uses examples drawn from the territorialization of North America, strategies employed by the church and the partitioning of space in the home and the workplace to highlight the role of territoriality as a component of power. He concludes that it is 'a device to create and maintain much of the geographic context through which we experience the world and give it meaning' (Sack 1986: 219). Territories are human creations, produced under particular circumstances and designed to serve specific ends. Once these territories have been produced, they become the spatial containers within which people are socialized.

---

from the territory. Territoriality is, thus, 'a primary geographical expression of social power' (Sack 1986: 5). The creation of states and the contingent drawing of boundaries between neighbouring states represent the most obvious political expression of territoriality. Power is being exercised over that particular bounded space through systems of rules which govern entrance and behaviour. In this way, territoriality can be seen as the spatial form of power. As highlighted by Taylor, 'territoriality is a form of behaviour that uses a bounded space, a territory, for securing a particular outcome' (1994: 151).

However, it is not just overt political power that can be expressed territorially. Sack uses examples from a variety of spatial scales. Thus, a street gang claiming space is a territorial reflection of the assertion of power. The demarcation of particular rooms within a house or workplace is also a form of territorial behaviour through which power (the power to include or the power to exclude) is expressed. For example, the prohibition of employees from certain rooms or areas within their place of work is an assertion of managerial power. In a similar vein, parental power is often expressed territorially through, for example, prohibitions on children entering some rooms in the house such as the kitchen or the parents' bedroom. Our need for personal space can also be seen as a form of territorial behaviour, the desire to have our own portable micro-territory (Hall 1959).

The key point of Sack's argument is that people behave territorially because they need to, or perceive the need to, not because it is an innate characteristic.

Following from Sack, Manzoni and Pagnini (1996) have identified the two principal ingredients underpinning territoriality – namely space and power. Using the example of Antarctica they suggest that, while space operates (Antarctica occupies physical space), power operates only weakly in this context. Thus, they argue that Antarctica is only a metaterritory because power is not rigidly expressed in the sense of actively enforced territorial claims.

Sack's work is important in not only emphasizing the political context of territorial behaviour but also highlighting how territoriality as a strategy operates at all spatial scales from the geopolitical strategies of superpowers down to the home and the workplace. More significantly, Sack identifies a number of what he terms 'tendencies of territoriality'. Chief among these is the fact that territoriality involves a classification by geographic area. Space can be apportioned between states or between individuals – this room or office or desk is *mine*, not *yours*. Second, territoriality is easy to communicate via the use of boundaries. These boundaries indicate territorial control and, hence, power over prescribed space. Third, territoriality is also a means through which power is reified. Through the visibility of land (or a room or desk) power can be 'seen'. Perhaps most importantly, through this reification of territoriality, attention is deflected from the power relationship. In this way, the focus is on the territory, not on the controlled and the controller. As Sack observes, 'territory appears as the agent doing the controlling' (1986: 33). As an encapsulation of this he notes the use of the term 'the law of the land'; we obey laws enacted in the name of a territorial formation rather than a set of people in power.

It is also worth pointing out that territory does not need to be intimately known in order for claims to be expressed. Laying claim to uncharted lands was commonplace during the era of imperialism, leading to often unpredictable long-term consequences. A classic example of this was the Treaty of

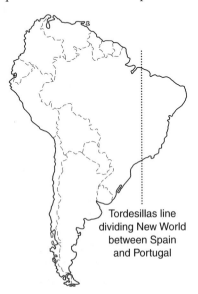

Tordesillas line
dividing New World
between Spain
and Portugal

Figure 2.2   Tordesillas line.

Chapter 3

The territorial state

The division of the world into bounded political units, commonly referred to as states, is the best-known example of formalized territories. Essentially the countries which form the world political map are states. Quite often the terms 'nation' and 'state' are used interchangeably, while contemporary states are often referred to as nation-states. This tendency serves to confuse rather than to clarify. A state is a legal and political organization with power over its citizens, those people living within its boundaries. A nation is more nebulous. It is a collection of people bound together by some sense of solidarity, common culture and shared history. Usually, this sense of common identity is underpinned by a historical attachment to a particular territory or national homeland. In some instances there may be a close approximation between nation and state. In France or Japan, for example, the vast majority of the state's inhabitants would see themselves as French or Japanese. The term nation-state serves to provide an impression of national and cultural homogeneity within the borders of a given state. However, all states contain within their borders a variety of nationalities. While the nation refers to a social collectivity, the state refers to a set of political institutions which have jurisdiction over a specified territory. The concept of the nation, and its importance in contributing to state formation and territoriality, together with the associated ideology of nationalism are considered in Chapters 4 and 5. This chapter focuses on the state and its territorial dimensions.

The state is currently the world's dominant form of political organization. When we look at a political map of the world what we see is a division of territory between these political units (see Figure 1.1). It is tempting to view this as 'natural' and people's everyday thinking is heavily permeated by the 'taken for granted' presence of these states. In the upheavals following the fall of the Berlin Wall and the break-up of the Soviet Union in the late 1980s and early 1990s many people, including many political analysts, were taken by surprise. In part, this reflects the extent to which it is sometimes assumed that states are immutable. We tend to have a very state-centred view of the world. In recent years the number of states has risen dramatically. In 1930 there were only about seventy; now there are approximately 200. The collapse of communism and the resultant break-up of the Soviet Union, Yugoslavia and Czechoslovakia led to the creation (or, in some instances, re-creation) of a number of new states such as Kazakhstan, Ukraine, Slovenia (Figure 3.1). As a consequence, a total

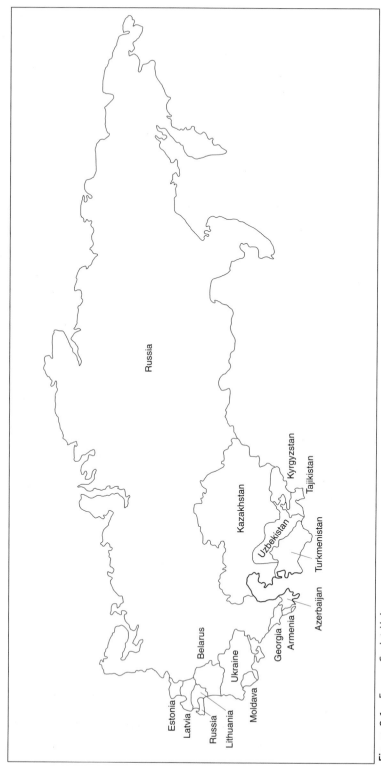

Figure 3.1   Former Soviet Union.

of twenty-two states replaced the previous three. This process of political and territorial reconstitution has re-opened a number of debates concerning national and ethnic identity and the location of borders in eastern Europe.

This chapter examines the state as a territorial unit and as a political unit. It explores theories of the origins of states before discussing the key features of states. Theoretical overviews of the role and functions of the state in the contemporary world are subsequently examined. The chapter concludes by exploring the issue of citizenship, the relationship between the state and those living within its borders.

Before proceeding, however, it is also important to distinguish between state and government. The state is a continuing apparatus of power while governments are the agents which carry out the day-to-day running of the state. The Labour Party is not the United Kingdom; it is merely the political party which operates, at the time of writing, the state known as the United Kingdom. The argument might be made that in totalitarian regimes there is an overlap between government and state where the ruling party essentially constitutes the state. However, most of the states of eastern Europe have continued to exist after the demise of their one-party Communist governments, notwithstanding the break-up of Czechoslovakia, Yugoslavia and the Soviet Union itself. These states have outlived their one-time rulers.

Origins of states

As suggested earlier, states are not 'natural' entities. They are human creations. They represent a formalized division of the world into political units. As such, they have not always existed. Prior to coming into existence in their current form, early examples of states can be found throughout history. The best known of these include Ancient Egypt, Greece and Rome. Greece consisted of a series of city-states while Rome was the centre for a spatially extensive empire. In other parts of the world state equivalents also existed. Examples include the empires of the Toltec, Aztecs, Mayans and Incas (Figure 3.2). Such territorial units waxed and waned. However, the concern here is with the origins of the modern state system. Geographers and others have been instrumental in developing theories of state formation and evolution. Two principal types of theory can be identified here – developmentalist and functionalist.

Developmentalist theories

These essentially view the process of state formation as one of territorial accretion. The German geographer Friedrich Ratzel proposed the idea of the state as an organism. He argued that states would expand territorially outwards in order to increase their size and power. In so doing they would devour smaller surrounding states (see Box 3.1). This serves as an excellent example of nineteenth-century geographers using what were biologically based ideas of territoriality (referred to in Chapter 2). State expansion was the manifestation of a struggle for space (Muir 1997).

Figure 3.2 'Proto-states' in central and south America.
Source: Glassner, 1993

Ratzel appears to have based his theory of state expansion and territorial control on Herbert Spencer's idea that human societies are social organisms. For Ratzel, every organism needed territory; human societies, as organisms, required territory. Population growth led to pressures which required the acquisition of more territory or extra living space – *lebensraum*. Ratzel's ideas accord with those of some nineteenth-century thinkers, such as Friedrich List and John Stuart Mill, who argued that the ability of a territorially based community to attain or maintain nation status was dependent on its size in population terms (Hobsbawm 1992). It followed that a large population and an extensive geographical coverage were necessary prerequisites for statehood. Given the increasing fragmentation along national lines currently occurring, and the longevity of entities such as Luxembourg, theories of the stability of states need to take into consideration other factors besides size.

These ideas espousing the perceived 'natural' or organic need for more territory served as a useful justification for European colonial expansion. The need for territory was a convenient intellectual justification for overseas conquest. This connection between geographical theory and practical politics is drawn out by a number of authors. The extent to which geography is implicated in Europe's colonial past demonstrates that geographical and social

Box 3.1   Friedrich Ratzel

Friedrich Ratzel was a German geographer who wrote an important series of geography texts in the nineteenth century. His ideas on the territorial expansion of states are linked to a school of thought known as environmental determinism. Essentially, this proposed that human behaviour was largely determined by the natural environment. Thus, climate and topography were seen as major influences on the way humans behaved. It is now widely accepted that people's behaviour, while it may well be affected by elements of the natural environment, is certainly not determined by them. Humans are subject to stimuli within the social environment and much of our behaviour is socially conditioned rather than biologically or naturally determined. It is also apparent that environmental determinism tended to bolster forms of scientific racism whereby people from certain places were judged to be inferior (Peet 1998).

Ratzel was also an ardent supporter of the German state and his intellectual ideas merged with his political views. He argued that states behaved as organisms and as such they needed to acquire more territory in order to remain strong and survive. He coined the term *lebensraum* to denote the idea of extra living space required by the state. Although he was dead before Hitler came to power, the idea of *lebensraum* was elaborated on by others and it eventually became central to Nazi ideology, helping to engender support for German territorial expansionism in the 1930s, ultimately resulting in the Second World War.

Table 3.1   Van Valkenburg's evolutionary theory of the state

Youth	US 1776–1803
Adolescence	US 1803–1918 – expansion
Maturity	1918 onwards – peaceful quest for international co-operation
Old age	Many European states

Darwinist ideas were useful in providing a rationale for Britain and other European powers to engage in the acquisition of territories which have now become the present-day states of Africa, Asia and the Americas. Geography and geographers, such as Halford Mackinder and Karl Haushofer, were actively implicated in this imperial project (Livingstone 1992; Ó Tuathail 1996). Indirectly at least, Ratzel's ideas were incorporated into Hitler's expansionist idea of *lebensraum*, an intellectual construct which formed part of Nazi ideology thereby contributing to the outbreak of the Second World War.

Another developmentalist theory, drawing on analogies from nature, is one proposed by van Valkenburg (1939). In this, states are assumed to evolve through a series of stages. Van Valkenburg's theory mirrors W.M. Davis' theory of rivers, familiar to physical geographers. He proposed a four-stage trajectory from youth, through adolescence and maturity, and ultimately to old age (Table 3.1). Leaving aside the somewhat benign view of the United States, this model endeavours to show how states are seen to evolve over time.

It assumes a positive linear trajectory whereby the state becomes stronger, more stable and more resilient.

Functionalist theories

State formation can also be viewed in functionalist terms. These ideas are associated with the geographer Richard Hartshorne (1950), who argued that states are under two types of pressure (Table 3.2). The first of these is centrifugal forces. These are elements which act so as to pull apart or destabilize the state. These factors might include a larger land area. It is assumed that the larger the territory covered the more difficult it is to retain control of it. Clearly, however, whether a state maintains its territorial integrity cannot be a function merely of size. It will depend on human features within its borders. Elements such as the extent of regional inequalities, ethnic diversity, language differences, etc., will be influential. In a state with sizeable regional differences, the poorer area may feel that it is being exploited by the richer and may, therefore, wish to secede on the basis that it would be better off 'going it alone'. Alternatively, the richer region may view the poorer region as a drain on its resources. The end result, again, might be secession. In part, Scottish devolution revolves around arguments as to whether Scotland would be economically better off alone, combined with a feeling that the benefits from Scotland's resources, most notably North Sea oil, have been siphoned off by England.

Table 3.2   Pressures on states

Centrifugal	Centripetal
Large land area	Small land area
Regional disparities	Spatially equitable
Ethnically or culturally diverse	Ethnically or culturally homogenous
Linguistic differences	Common language

The ethnic diversity of the population will be a key element in determining the stability of the state. It is suggested that ethnically diverse states are more likely to be unstable. However, this is only the case if a particular ethnic group feels that it has been discriminated against and has been 'politicized' to the extent that it wishes to attain nationhood and thereby acquire its own political apparatus, namely statehood. Linguistic or religious differences (often associated with ethnic differences) can lead to pressures for separation. Currently, Belgium's continued existence is being called into question. Here the principal fault line is language with a division between Flemish speakers and French speakers (Walloons) (Figure 3.3). However, the example of Switzerland as a stable multilingual democracy (there are four official languages) suggests that language differences do not necessarily create political instability. The key element here is whether particular groups see themselves as constituting a cohesive body such that they feel themselves to be a 'nation' entitled to their own state. In general terms, it is argued, the closer the coincidence between nation and state, the more stable the state.

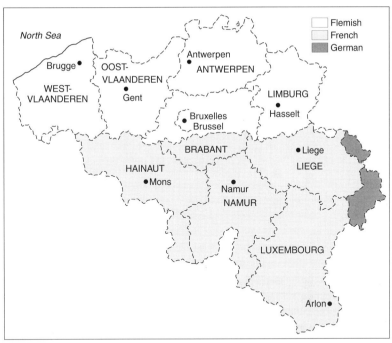

Figure 3.3 Linguistic and provincial divisions in Belgium.
Source: Murphy, 1995

The second type of pressure to which states may be subject is centripetal forces. These serve to bind the state together. If centrifugal forces act as sources of instability, then centripetal forces have the opposite effect. They might include elements such as a common language, religious homogeneity, lack of extreme regional inequalities, etc. Under such circumstances, it might be argued that there will be few pressures on the state as currently constituted.

Interstate system

There are a number of difficulties with the types of model outlined above. Both developmentalist and functionalist models are classics of ex-post rationalization. Underpinning these theories is the idea that states expanded out from original core areas (Figure 3.4). What makes this inevitable? What alternative cores might have developed outwards but failed to do so? Was France inevitable? How many alternative states to our current ones might have existed? Such theories take the existence of the present-day state system for granted. Both developmentalist theories and functionalist theories are based on existing political realities. As such, they are premised on the basis of what actually happened rather than what might have happened and they consequently fail to explain why particular political configurations arose rather than others. Each state's initial existence is not subject to question. It is merely its size and stability which are up for consideration. Such theories are examples of 'reading history backwards'.

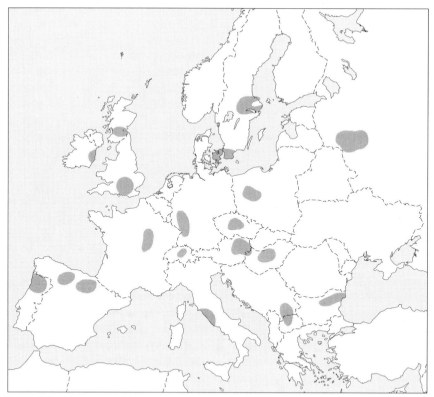

Figure 3.4   Core areas of European states.
Source: Taylor and Flint, 2000

Developmentalist theories, like most 'stagist' theories, do not take account of other states. They ignore the fact that one state's territorial expansion is at the expense of one or more other states. Not all can develop along the lines proposed. The very fact of their acting out this scenario means that another state, or potential state, is failing to do so. Again, like other similar theories, such as Rostow's stages of economic growth, it is also a European model. The empirical basis is derived largely from the European experience which has then been projected onto the non-European world.

Developmentalist and functionalist theories tell us something about the evolution of states but the explanations offered are narrow. It is certainly the case that regional and ethnic tensions can have a significant bearing on the stability of a state, but, rather than a narrow focus on internal characteristics, relative power balances between states need to be taken into account when attempting to understand the evolution of a world state system. Taylor and Flint (2000) suggest that Wallerstein's world systems theory allows state forma-tion to be viewed from the 'outside' through the lens of interstate relations rather than from the 'inside' with its reliance on states' internal characteristics. Such a perspective also allows us to see states as elements within a broader capitalist system, rather than viewing them in isolation. Conventional views of the state tend to view it as a set of institutions freely entered into. Clearly, states can be imposed on people. Much state-building in Africa originates with an external imperial power creating a territorially bounded unit to serve its own interests (Warner 1998). Issues such as contested sovereignty are thus of vital importance. In addition, competing nationalisms (which are not necessarily issues internal to the borders of existing states) need to be looked at more fully. Nations and nationalism are discussed in detail in the next chapter.

Key features of states

States are spatial entities comprising four essential features:

- territory
- people
- boundaries
- sovereignty

Territory and people

That a state must possess territory and people is largely self-evident. While the state is a political organization, it must have a territory, a portion of geographic space (land, air, water) over which it exercises power. States are endowed with the power to legislate for their territory and for those people living in that territory. As Mann argues, 'only the state is inherently centralised over a delimited territory over which it has authoritative power' (1984: 69). This follows from Weber's theoretical formulation which identified the state as having a monopoly of legitimate force over a designated territory. Thus, the state can be seen in relation to four key elements – monopoly, territory, legitimacy and force. This means that it is different from transnational corporations and other institutions of power which are not spatially restricted in the same way.

Given that states are human creations, and that they are run by people who exercise power over the citizens of the state (and, in many instances, are assumed to exercise that power in the name of the citizens), it is impossible to have a state without any people. People constitute both the government and the governed.

Boundaries

In an interstate system, boundaries are necessary to separate states from each other. If states have control over designated territory, it follows that there must be recognized boundaries separating their own territory from that belonging to neighbouring states. Clearly, boundaries may be contested; they may change over time or they can disappear, as happens when two states merge. Here it is necessary to distinguish between boundaries and frontiers. Normally a boundary is taken to refer to a precise line separating one territory from the next, while a frontier is the zone around the border between two adjoining states. However, the terms are quite often used interchangeably.

Traditionally, political geographers have distinguished between 'natural' boundaries and 'artificial' boundaries. The former category encompassed naturally
occurring phenomena, such as rivers and mountain ranges. The Rio Grande
river forms part of the border between the United States and Mexico, while
the Mekong river forms part of the boundary between Laos and Thailand.
Lines of latitude are seen as examples of artificial boundaries. Thus, the 38th
parallel forms part of the border between North and South Korea, while the
49th parallel forms part of the border between the USA and Canada. However, the distinction between natural and artificial boundaries is somewhat
misleading given that borders, whether naturally occurring features or not, are
intrinsically artificial. States are human creations; it follows that their borders
are, therefore, artificial. As a consequence of political decisions, some rivers
become borders (or cease to be borders), others do not.

The more traditional geographers' concern with a classification of boundary types is of much less significance than an exploration of the impact of
the imposition of borders. Borders are constantly subject to change resulting
from disputes over territory. Disputes may relate to the precise location of the
border or they may centre on whether or not a particular border should exist
in the first place. Numerous examples of changing borders exist. Glassner
(1993) indicates the shifting boundaries of Bulgaria between 1878 and 1947
when its present-day borders were agreed on (Figure 3.5). The German–
Polish border has been subject to debate and the present one was only finally
accepted following German re-unification in 1991. The boundary between
France and Germany has also altered during the course of this century as
a consequence of both world wars, with part of Alsace-Lorraine switching
between countries. Similar changes occurred along the Belgian–German border.
Some borders are subject to on-going dispute such as those between Venezuela
and Guyana and Surinam's borders with both Guyana and French Guyana
(Figure 3.6). This is a legacy of a division of colonized territory between
European powers when the areas in question had not been fully 'explored' by
Europeans and, hence, were never accurately mapped.

While the precise location of a border may give rise to conflict, in other
instances the very existence of the border is what is in dispute. The political
conflict in Ireland revolves around the existence of the border between the
Republic and the North (Plate 3.1). For Irish republicans, who wish to see a
unified Ireland, the border symbolizes the conflict. It is seen as an imposed
and artificial boundary whose removal is essential. The border has been seen
by Unionists as offering a convenient escape route for members of the Irish
Republican Army (IRA) who are able to cross this permeable frontier to what
is seen as the 'haven' of the Republic. In this situation, the border has become
a key site within the conflict. It has been a heavily fortified area which has
witnessed many violent episodes throughout the conflict with attacks on British checkpoints and customs posts.

While borders are political constructs, they have clear social and cultural
implications, particularly for those living in border zones. Borders are not just
lines dividing territory; they are social and discursive constructs which have

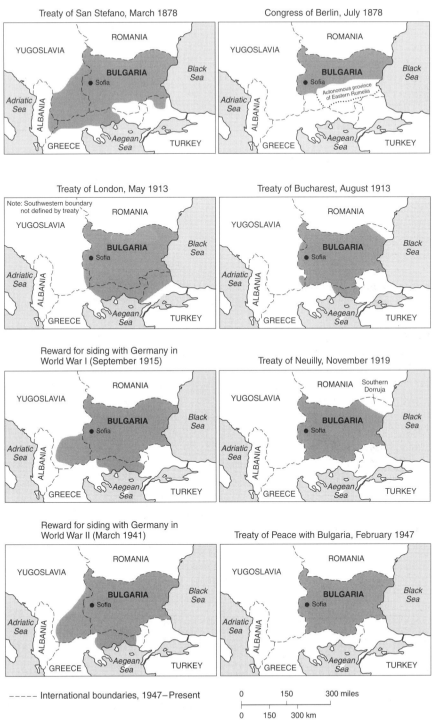

Figure 3.5 Bulgaria's shifting borders 1879–1947.
Source: Glassner, 1993

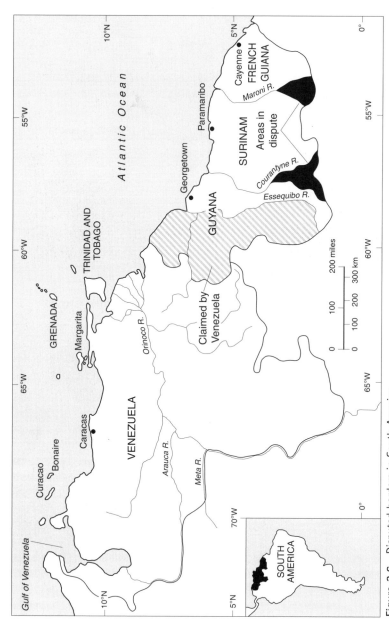

Figure 3.6 Disputed borders in South America.
Source: Glassner, 1993

Plate 3.1   Border between Northern Ireland and the Republic of Ireland.

important ramifications, not just in a broad political sense but also in people's everyday lives (Newman and Paasi 1998; Paasi 1999). Studies in various parts of the world, summarized by Prescott (1987), suggest that borders often cut through areas which are culturally homogenous, or at least in which there are no significant socio-cultural differences either side of the line. However, the creation and existence of formalized borders may lead to significant differences on opposite sides (Rumley and Minghi 1991). For example, differences in agricultural practices, levels of urbanization, income levels, etc., may emerge as a consequence of differing government policies. Contrasting landscapes may evolve in terms of phenomena such as different settlement patterns (Plate 3.2). People may also develop different attitudes towards the border depending on whether they see it in a positive light and/or depending on whether they see themselves as being on the 'right' side of it. In many respects, the creation or imposition of a physical border can result in the creation of a partitionist mentality through which people on opposite sides drift apart owing to their political separation. The evolution of a sense of difference between 'east' and 'west' Germans might be seen as a consequence of that country's political division between 1945 and 1991, a division marked by a heavily patrolled border exemplified by the construction of the Berlin Wall in 1961 (Plate 3.3). That borders have such an impact is, in part, the reason they are so often centres of conflict (see Box 3.2).

In a myriad of ways, borders become elements within people's everyday lives and shape their day-to-day being. The significance of borders stretches from the global importance attaching to the Berlin Wall through to the local

Plate 3.2   US–Mexico border, Tijuana.

Box 3.2   Finnish–Russian border

Finland gained independence in 1917, having been an autonomous region within Russia since 1809. Prior to that it had been part of Sweden. Following independence, Finland moved to secure its borders as a means of asserting its newly acquired separate identity. The firming up of its border with the Soviet Union was accompanied by a diminution of trade with that country and an increasingly westward orientation, although Finland was to remain a neutral country throughout the Cold War. Cross-border disputes arose during the Second World War and the present border was affirmed in 1947. Finland ceded parts of Karelia to the Soviet Union and, as a consequence, 420,000 people were resettled.

The impact of a closed border was to re-orient the borderlands inwards towards their own state rather than towards each other. Since 1989 the border has been re-opened and people cut off from their original homes are now free to visit them. However, the border, combined with Soviet policies, has effectively altered the nature of the place. While cross-border trade and visits have increased since the demise of the Soviet Union there are substantial levels of mistrust and suspicion about the other side. There have been calls in some quarters for the re-integration of the 'lost' areas into Finland, although no official territorial claim is being made.

Source: adapted from Paasi, 1999

Plate 3.3   Berlin wall.

but very real separation of people by a see-through wall separating Nogales
in Mexico from the town of the same name in Arizona (Verhovek 1997).
Recently the Russian town of Ivangorod found itself without a water supply.
It received its water from the neighbouring Estonian town of Narva. Failure
by Ivangorod to pay for its water led to the decision to cut the supply. Prior
to the dissolution of the Soviet Union, both towns were part of the same
polity but now they are separated (Figure 3.7). As Meek (1999) suggests, the
two towns are growing apart economically, socially and culturally despite
being separated only by a river.

   In addition, it is worth pointing out that for many the existence of state
boundaries is, in some senses at least, largely meaningless. Many nomadic
groups such as Bedouin people living in Iraq and Jordan tend to ignore what
to them are meaningless lines on a map. Similarly, gypsies in many European
countries regard state borders as of no great significance in terms of identity,
in that they may have no strong allegiance to the state in which they live. This
is not to suggest that borders are of no significance to these groups in other
ways. Their treatment may be better in some states than in others while the
existence of political borders has consequences for their ability to move freely
between states (Fonseca 1996).

Sovereignty

In addition to land, borders and people, there is one vitally important
additional feature which states must possess: sovereignty. This refers to the
authority of a state to rule over its territory and the people within its borders:
that is, the right of the state to rule without external interference. Such interference (invasion, etc.) is seen to be a transgression of international law,

Figure 3.7   Location of Narva and Ivangorod.

as many have argued in relation to NATO air attacks on the Federal Republic
of Yugoslavia in 1999. People may have territory but not sovereignty. This
feeds into a second important implication of sovereignty. A state must be
recognised by other states. In an interstate system, if a particular state's sovereignty is not recognized, then its legitimacy is called into question.

   Sovereignty can only occur through external recognition. During the apartheid era in South Africa a number of *bantustans* were created by the white
minority government. These were ostensibly devised as self-governing 'homelands' for black residents in what was essentially a charade designed to imply
the existence of political autonomy for sections of the black population. These
homelands, pronounced 'independent' by the South African state, were not
sovereign because the international community refused to recognize them as
such. Only South Africa viewed them (or claimed to view them) in this way.
(This issue is returned to in Chapter 7.) A good contemporary example is the
so-called 'Turkish Republic of Northern Cyprus'. This was created following
a Turkish invasion of Cyprus in 1973 as an apparent response to oppression
of Turkish Cypriots by the majority Greek Cypriots. However, only Turkey
recognizes this 'state' (Figure 3.8).

   The break-up of the former Yugoslavia and the creation of its successor
states were sanctioned by the international community through the recognition of the sovereign claims of the newly created (or re-created) states. However,
this recognition of sovereignty has of course been contested. Many Serbs have
been reluctant to concede territory to any of the non-Serb states, particularly
Bosnia-Herzegovina and Croatia. Staying in the former Yugoslavia, another
interesting example of contested sovereignty concerns the entity internationally
(at least officially) referred to as 'Former Yugoslav Republic of Macedonia'.

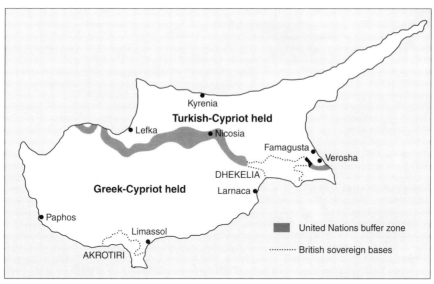

Figure 3.8 Partition of Cyprus.
Source: Kliot and Mansfield, 1997

The problem here arises from Greece's concern that full recognition of an entity known as Macedonia will lead to claims for an enlarged state incorporating part of northern Greece, which has a province called Macedonia. At present the country is listed in atlases and gazetteers as 'FYROM'.

In most instances claims over disputed territory are linked to nationalism, as explored in more detail in the following chapter. Examples of long-running territorial disputes include Morocco's claim to Western Sahara, which is not officially recognized. Iraq's annexation of Kuwait in 1991 and the annexation of East Timor by Indonesia in 1976 represent overt examples of territorial claims, leading in the case of Kuwait to a major international conflict resulting in the deaths of many combatants and non-combatants and in the instance of East Timor to the massacre of many thousands of people. At the time of writing, violent conflict continues in the latter in the wake of a referendum in which the overwhelming majority of East Timorese voted in favour of complete independence from Indonesia.

Questions may arise as to how many other countries must recognize the sovereignty of another for it to acquire 'recognized' status. Clearly, some countries may perpetually refuse to accord full recognition to the existence of others. Following the unification of Vietnam in 1975, after its war with the United States, it was many years until some countries recognized it as a sovereign entity. Indeed, the USA and UK did not do so until some twenty years later. The Federal Republic of Germany (West Germany) refused for many years to recognize the sovereignty of the German Democratic Republic (East Germany).

Sovereignty disputes interconnect with border disputes. The African state of Eritrea attained independence in 1993 following a protracted thirty-year

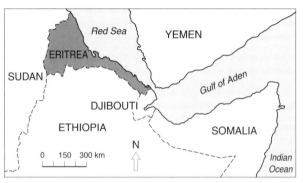

Figure 3.9   Eritrea.

struggle to secede from Ethiopia. Currently, the border between the two is disputed, with both countries claiming sovereignty over stretches of territory. In fact, Eritrea has border disputes with all of its neighbouring countries (Figure 3.9). A number of Yemeni soldiers were killed in 1998 in a dispute with Eritrea over control of islands in the Red Sea, over which both countries claim jurisdiction. In South America, Ecuador has a long-standing claim to sovereignty over part of northern Peru (Radcliffe 1998).

A number of spatial entities have existed which might be said to exhibit, in part, the characteristics of states. The most obvious examples are colonies of imperial powers. While these might have had recognized boundaries, they were not independent sovereign entities. They were politically dependent on the external power. The remnants of colonialism still exist. Until its reversion to Chinese rule in 1997, Hong Kong was a British dependent territory. It was not a sovereign state. Britain still has a number of colonies, although the word colony is no longer used to describe them. In the 1950s the term 'Dependent Territories' was introduced; recently this has been replaced with the term 'British Overseas Territories'. These are small entities, often islands, which have not as yet achieved full political independence from their colonizing power. Between 1922 and 1971 Northern Ireland had its own parliament. However, it was not a sovereign political entity (nor did most of its residents wish it to be) as responsibility for things such as foreign policy resided with the UK government. In addition the 'statelet' was (and still is) heavily sub-sidized from London.

Interestingly, despite their name, the individual states of the USA are not sovereign independent entities. They are elements within a federated structure. Sovereignty resides with the United States of America. The same applied in the case of the former Soviet Union. Despite its title of Union of Soviet Socialist Republics, these republics were anything but sovereign (and far from socialist, many would argue). German *länder* have considerable autonomy, and their own regional governments, but they are not sovereign entities.

This discussion has highlighted the nature of territoriality at the level of the state. There is, however, a need to be wary of not falling into what Agnew (1994) refers to as the 'territorial trap', whereby states are assumed to be immutable entities. Many discussions seem to suggest, boundary changes and disputes over sovereignty notwithstanding, that states will continue to exist. This state-centred approach reifies current political–territorial structures and tends to ignore the fact that, rather than being permanent features, they are historically contingent.

It is important to bear in mind that states are more than mere spatial containers. It should be obvious that, while states are spatially bounded entities, they are not static. States attempt to expand, to take over other territory; they colonize, they engage in conflict with other states. As suggested in Chapter 2, geographers have argued for the treatment of place as something more than an external 'commodity'. We live our lives in places and we imbue those places with meaning. Shared experience of places provides a basis for communal identity. A collective consciousness develops out of shared residency. Places are dynamic and people re-shape the places in which they live. In exactly the same way, states can be seen as dynamic entities, shaped and re-shaped through the interactions of human political activity.

The role and functions of the state

Up to this point aspects of state formation and conflict between states have been explored. This reflects the idea of the state as a geographical entity. However, the state is primarily a political unit and questions arise concerning what it is that states do. What is the function of the state? At its most basic level, it can be said that the state provides a legal framework, infrastructure and services to be used for the benefit of its citizens. The state does the following:

1. It regulates the economy (although currently dominant economic theories suggest that states should minimize the exercise of this function).
2. It provides public goods such as health and transport services (although there is a contemporary trend in many countries of privatizing such services).
3. It provides legal and other frameworks which guide its citizens' behaviour.
4. It defends its territory and its people against external aggression and internal threats.

People come into regular contact with the state. In most countries letters are delivered by state-employed postal workers. Telephone and transport services are usually run by the state or by semi-state companies. Streets are policed by a state police force. Many utilities are provided by the state. In this way, there is a series of daily bonds between the state and its citizens. The state becomes a routinized element in people's day-to-day existence.

In the United Kingdom since 1979, and in many other countries, there has been a policy of de-nationalization whereby many state services have been privatized. In the UK, services such as water supply and the railways are now operated by private companies rather than by the state. This policy of 'rolling back the state' can be said to have had a definite effect in reducing the amount of direct contact people have with 'agents' of the state. Despite this, however, the state remains quite pervasive. It needs to be borne in mind that even privatized utilities operate within a state-imposed regulatory framework. Even in the so-called private domain, the state is virtually omnipresent, legislating for the age of sexual consent, the age at which alcohol can be consumed, registering births, marriages and deaths and having a role in other aspects of peoples' lives. Thus, although the provision of many services has been privatized, the state remains highly interventionist, maintaining a highly regulated framework. In these various ways, Mann (1984) argues, civil society becomes territorialized. Conflict between groups takes place on a territorial plane, i.e. within the territorial framework of the state. The extent to which the state is, or should be, involved in service provision and regulation is hotly contested. On the right of the political spectrum it is argued that the state should interfere as little as possible with people's everyday lives (although this argument tends to relate primarily to welfare and service provision rather than to support for the private sector). On the left, it is generally argued that the state should play a key role in protecting its citizens and in regulating the territorial framework in which they live so as to ensure some degree of equality of treatment and protection for more vulnerable groups.

Theories of the state

There are many different types of state. There are unitary states where power is highly centralized and there are also federal states with a high degree of political devolution. Similarly, there are authoritarian states with quite rigid control and little popular participation, while there are more open state systems, symbolized for many by the liberal democracies characteristic of western Europe. At a simplistic level there is a tendency to think of the relationship between the state and its citizens as a freely entered-into set of arrangements. Clearly, in authoritarian states this is most definitely not the case. However, even in so-called liberal democracies, in which electoral systems are reasonably free and transparent, there is a need to delve beneath the surface appearance of the state as the repository of the collective will of its citizens in order to devise adequate theories of the role and functions of the state.

This section summarizes the main types of theory relating to the state in liberal democracies. In broad terms, there are three main bodies of theory relating to the state. These are pluralist theories, elite theories and Marxist theories (Figure 3.10). More attention is devoted to the last of these as their particular strength is their usefulness as a critique of the other two.

Pluralist theories

From a pluralist perspective the state within liberal democracies is viewed as a neutral arbiter. It is seen as being above, and separate from, any vested interest. It is deemed to be apolitical in the sense of having no interest in the form of society. Rather, it is seen as an institution which is shaped according to the citizens' will, democratically expressed. This reflects the basic democratic

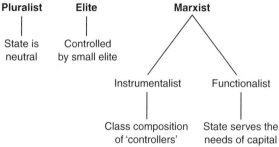

Figure 3.10 Theories of the state.

underpinnings of liberal democracy. People vote in elections and sovereignty is seen to reside with the citizens. The notion of popular sovereignty, whereby power emanates (via the electoral process) from the people, derives from the philosophy of Hobbes and Locke. People are deemed to be active participants in the process of societal regulation. From this, it follows that governments merely act as the agents of the electorate. People transfer their power to their elected representatives who then act in their name and order affairs of state accordingly.

The classic liberal view of the state is that of the guarantor of the rights of the individual. This rests on the notion of a clear private sphere which should be protected from outside interference. The state, or public sphere, is necessary in order to safeguard this private sphere. The irony is that very often the private sphere is seen as being penetrated by the public sphere. In other words, the state is sometimes perceived as 'a potential threat to the liberty it exists to secure' (Schwarzmantel 1994: 42). Liberal democratic states attempt to reconcile the classic liberal position of non-interference and freedom with the democratic ideal of popular sovereignty.

Pluralist states are seen as the polar opposites of totalitarian ones. Instead of one imposed view, there is a reconciliation of a variety of diverse views. Although at any one time a particular political party may hold power, and although they may represent a particular vested interest, the nature of the pluralist state is such as to ensure a system of checks and balances which will safeguard the citizenry against abuses of power. In this way, an independent judiciary, media, etc., are seen as integral components of the state apparatus. No one faction can have total control. States, on balance, are seen as neutral and as responsive to the various pressures which emerge within society. The various elements of the broad democratic system are deemed to bring about appropriate solutions to the various conflicts which arise. The state merely delivers the policies and services needed by society. The state is thus seen in a technical rather than political sense.

Elite theories

These derive from the ideas of people such as Gaetano Mosca (1939) who suggests there is a distinction between a ruling class and a class that is ruled.

The argument here is that there will always be a stratum in society who will tend to rule in the interests of the rulers rather than the ruled. Self-interest will take precedence over any broader concerns. This rests on a sociological distinction between an 'elite' on the one hand and a 'mass' on the other. In a variant on this, Joseph Schumpeter (1976) has argued that people choose between competing elites. In this way there is not, strictly speaking, a pluralist state but rather a constrained choice between various aspirants to power. This has been termed democratic elitism and, in essence, it forms a bridge between Marxist theories and pluralist theories. Critics of elite theory argue that it is predicated on the idea of an innate tendency towards elite formation. Why do elites necessarily exist? Is this somehow an intrinsic part of human nature? The idea that elites are 'natural' is subject to the same criticisms as theories of the innateness of territoriality discussed in Chapter 2.

Marxist theories

Marxist theories of the state are predicated on the idea that societies are divided along class lines. There exists a capitalist class (those who control the means of production, in Marxist parlance) and a working class (or proletariat) who sell their labour to the capitalist class. The surplus value (profit, essentially) gained by the capitalist class as a consequence of employing the workers is seen as resulting from the exploitative nature of the relationship between these two classes. While most Marxist theoreticians would agree that the contemporary world is more complex than in Marx's day, nevertheless the basic foundations of the analysis remain the same. The distinction between a powerful class and a relatively powerless one is the basic fracture in capitalist society.

Marxist theorists argue that both elite theory and pluralist theory essentially deal with surface appearances. They argue that what is important are the deep-seated structures within society; the hidden dimensions of power. Class structure is seen as the key element. The state is seen as a mechanism which acts in defence of the class structure. In this way, rather than the state being neutral, it defends the interests of the dominant class.

From a Marxist perspective the state cannot be neutral. Rather than being seen as a neutral mediating party between various vested interests, it is itself a vested interest. It acts in the interests of the prevailing capitalist system. Viewed this way, the state is not an independent structure standing apart from the rest of society. On the contrary, the state is deeply embedded within socio-economic relations (Held 1989). Marx himself did not devise a theory of the state. In *The Communist Manifesto* of 1848, Marx and Engels saw the state as 'a committee for organising the affairs of the bourgeoisie'. In other words, in orthodox Marxist terms the state is seen as an agent of the ruling class. It is, therefore, seen as being necessary to deal with capitalist crises. It defends and protects the class structure by preserving existing power imbalances. In the words of Engels, 'the state is an organization of the possessing class for its protection against the non-possessing class' (1884; reproduced in Held *et al.* 1985: 105).

Capitalist Society (Quartet, London, 1969) and the responses by Nicos Poulantzas, particularly 'The problem of the capitalist state', *New Left Review* 1969, **58**: 119–33.
Useful overviews of the issues surrounding citizenship include J.M. Barbalet, *Citizenship. Rights, Struggle and Class Inequality* (Open University Press, Milton Keynes, 1988) and Maurice Roche, *Rethinking Citizenship. Welfare, Ideology and Change in Modern Society* (Polity Press, Cambridge, 1992). Changing ideas of citizenship in the context of European integration are discussed by Paul Close in *Citizenship, Europe and Change* (Macmillan, Basingstoke, 1995) and *Citizenship, Nationality and Migration in Europe* edited by D. Cesarani and M. Fulbrook (Routledge, London, 1996). For a more intricate discussion see Rainer Baubock, *Transnational Citizenship. Membership and Rights in International Migration* (Edward Elgar, Aldershot, 1994). See also the special edition of *Political Geography* 1995, **14**(2) on 'Spaces of citizenship'.

Chapter 4

Nations and nationalism

In the previous chapter the nature and functions of the state, the primary building block in the political map of the world, were explored. This chapter examines the concepts of nation and nationalism, key elements in the creating and sustaining of a world of states. We live in a world where the existence of nations, like states, is taken for granted. Nation, nationality and nationalism are terms used regularly in the media and in everyday discourse. National identity is refracted through the singing of national anthems, support for national sports teams and in a variety of other ways. The 'taken for granted' existence of nations and the associated political ideology of nationalism underpin the configuration of the world political map. Many of the disputes between countries and, indeed, many of those occurring within the borders of certain countries centre around competing nationalisms. In most instances disputes between national groups are concerned with claims to territory. In many cases these competing claims lead to extremely violent conflict, as in the former Yugoslavia, Northern Ireland and Israel–Palestine.

While such overt conflicts tend to make us aware of the importance of nations and nationalism, there are many other, perhaps more subtle, ways in which national differences and, hence, the existence of nations are manifested. Differences can be observed in dress, food, language and a variety of other things. Even a sense of national names exists. Thus, Kelly and Murphy are assumed to be Irish, Perez Spanish, Müller German, and so on. These serve to reinforce our sense of a world composed of nations. In the words of Michael Billig, 'the world of nations is the everyday world' (1995: 6). This chapter focuses on ideas of national identity and discusses the origins of nations, leading to a consideration of the territorial ideology of nationalism. Different forms of nationalism are outlined and the debate over the positive and negative implications of nationalism is considered. Finally, what can be regarded as the functions of nationalism are examined.

Nation and state

To reiterate the point made in the previous chapter, nations and states are two distinct conceptual entities. The latter are agencies with power over citizens within demarcated territory while the former are more nebulous. Nations are social collectivities with an attachment to a certain territory. Hugh Seton-Watson

Figure 4.1 Kurdistan.

suggests that 'a nation exists when a significant number of people in a community consider themselves to form a nation, or behave as if they formed one' (1977: 5). It follows from this that, to a large extent, a nation is a mental construct as much as a physical reality. More concretely, Anthony D. Smith defines a nation as 'a named human population sharing an historic territory, common myths and historical memories, a mass public culture, a common economy and common legal rights and duties for all members' (1991: 14). The last two elements imply the achievement of nation-state status, not something attained by all people claiming to be a nation.

Despite the differences between the two, nation and state are obviously closely interconnected. This relationship is succinctly described by David Miller: ' "nation" must refer to a community of people with an *aspiration* to be politically self-determining, and state must refer to the set of political institutions that they may aspire to possess for themselves' (1995: 19, italics in original).

It is obvious that in some instances nations and states closely coincide. France is typically cited as an example of this – a state peopled mainly by those who regard themselves as French. However, such a view may have important implications for those perceived as 'non-French' but residing within the national territory. While France might be regarded as a close approximation to the idealized nation-state, where a territory is marked by a coincidence of national homogeneity and political control, there are many within France, particularly Algerians or Moroccans, who may suffer racial discrimination as a consequence of not being seen to properly belong to the nation. There are also those, such as Bretons and Basques, who may not necessarily see themselves as primarily French.

Despite much discussion of rights to national self-determination, there are many nations which do not have a state. One of the more notable examples is that of the Kurds, most of whom live in Iraq, Turkey and Syria, countries in which, by and large, they are treated as second-class citizens (Figure 4.1). The aspiration towards an independent Kurdistan has not as yet been realized. Similarly, there are many states which encompass a number of nations, or peoples who would regard themselves as possessing a distinct national identity. The former Soviet Union and the former Yugoslavia are good examples, embracing as they did a wide variety of national groupings. Another good example of the mismatch between state and nation is the United Kingdom, incorporating, as it does, the nations of England, Wales, Scotland and part of Ireland. (The peculiarities of nationalism within the UK will be returned to in Chapter 5.)

There are also examples of nations which have formed a number of states. Reference is often made to the 'Arab nation', yet there are a number of Arab states such as Saudi Arabia, Iraq, Kuwait and Egypt. Equally, it is obvious that many states contain minority national or ethnic groups. Examples include aborigines in Australia and 'native Americans' in the United States. Similarly, there are states where one or more groups wish to secede in order to create their own state to reflect what they see as their right to full nationhood. This is the case with Basque separatists in Spain and Scottish and Welsh nationalists in the United Kingdom. The case of Northern Ireland is a specific form of secessionist nationalism in that those who wish to secede are seeking unification with, or incorporation into, another already existing state, the Republic of Ireland. This is known as irredentism.

Given the contemporary importance of nationalism and national identity and the strong emotions which they evoke, whether in outright war or in sporting or cultural contexts, it is important to have some understanding of the origins of nations and the evolution of nationalism as a political ideology.

The origins of nations

There is a tendency to think of nations as fixed, static entities which have always existed. It is taken for granted that we are born into a national group which has a historic attachment to a particular territory. However, this is somewhat simplistic. How did this national group come into existence? How and why did people come to perceive themselves as part of a collectivity of this nature? The existence of nations is not as straightforward as it might at first appear. There are a variety of theories concerning the origins of nations. For purposes of clarity, these theories can be subsumed within three main strands of thinking. The first are referred to as primordialist or essentialist theories. The second are perennialist theories and the third are modernist or situationalist theories. It is important to bear in mind that the inevitable reduction of quite complex arguments into broad categories results in oversimplification. However, this should still present a flavour of the nature of debate over nations and nationalism.

Primordialist theories

Primordialists see nations (as distinct from the more recent phenomenon of nationalism) as having quite deep-seated historical origins. Strict primordialists would argue that nations have always existed. This essentialist way of thinking sees nations as natural entities whose origins go back to time immemorial. Nations are seen as possessing some historical and immutable core. From this perspective, there is seen to be something which is essentially Dutch, or German or Serb. In this way of thinking ethnicity is seen as an extension of kinship

which, in socio-biological terms, is seen as the 'natural' social unit. An alternative approach is that language, religion, race, ethnicity and territory (components in the nationalist mix) are basic organizing principles. Viewed in
either terms, nations are seen as naturally occurring human phenomena.
This extreme primordialist approach is reflected in the idea of a 'pure' race or
nation. The words of an Irish folk song reflect a belief in something immutable at the core of the nation:

> Once upon a time there were
> Irish ways and Irish laws
> Villages of Irish blood
> Waking to the morning
> Waking to the morning

Successive invasions of Ireland, first by the Vikings and then by the English,
resulted in massacres and dispossession, yet:

> 800 years we have been down
> The secret of the water sound
> Has kept the spirit of a man
> Above the pain descending
> Above the pain descending
> (Moving Hearts, 1981)

Clearly there is a perception of something primordially Irish surviving all
these waves of invasion, an immutable purity underpinning Irishness derived
from a mythical pure past. In most instances attachment to territory is part of
this belief. Nationals may be deemed to have a deep bond with the territory of
the nation, revealed through references to 'soil'.

It should be obvious that such views are not only absurd but also potentially dangerous. They risk essentializing nationality and leading to witchhunts
against elements seen to dilute the essence of the nation. Hitler's assertion of
Aryan racial supremacy was predicated on this primordialist idea. Contemporary neo-Nazi organizations such as Le Front Nationale in France and the
British National Party in the United Kingdom have a version of this racialized
thinking at the core of their political philosophy (Figure 4.2). These parties
see their existence as one of defence of their nation against the dilution brought
about through such processes as unchecked immigration or interethnic mar-
riage. Such extreme right-wing philosophies exist in a number of European
countries, both east and west, at the moment and are exemplified in the
ethnic cleansing of Bosnia-Herzegovina, Croatia and Kosovo in the aftermath
of the dissolution of Yugoslavia.

Perennialist theories

A more nuanced and intellectually credible version of primordialist thinking
proffers a theory of national origins which accepts their constructed nature
but yet sees them as having a long historical lineage. Anthony D. Smith, while
rejecting essentialist ideas of the primacy of nations of the kind outlined
above, nevertheless sees nations as originating from what he terms *ethnies*

SAYING 'NO' TO A FEDERAL EUROPE
SAYING 'YES' TO BRITISH FREEDOM

ELECTION COMMUNICATION   WEST MIDLANDS REGION   JUNE 10th

"At last we've found a party which really cares about Britain and British values. A party worth supporting. The British National Party "

http://vote.bnp.net
☎ 01902 892769

IF YOU thought that all political parties are the same, that they all conspire to deny you a democratic choice on the really important issues, and that they all want your vote just so they can get their snouts in the trough — you were right! Until now!

Now there's a *new* party, one which is putting choice and democracy back into British politics; a party which is ignoring the dictates of political correctness and the anti-British 'spin' of the mass media. A party which is listening to what ordinary people really want. This is the British National Party and, as this election uses Proportional Representation, we stand *a real chance of winning* seats in the European Parliament. And when we win, *you* win, because our policies are what you've wanted to see for years. That makes sense, doesn't it? Don't take our word for it, have a look overleaf and judge for yourself!

Make a difference as you vote
BRITISH NATIONAL PARTY
SAVE OUR STERLING

Figure 4.2   BNP election leaflet. The reverse side of the leaflet refers to the need 'to preserve the cultural and ethnic identity of the British people' and calls for an end to immigration.

which were extensions of kin groups. These groupings developed feelings of community over time. Feelings of belonging together, of collective identity, of attachment to their locality would have evolved gradually. Smith argues that such feelings have waxed and waned with some ethnies becoming more self-conscious and dominant than others. Vertical suppression of subordinate ethnies would have taken place as the intelligentsia played a role in inculcating a common culture aided by processes such as the harmonization of vernacular languages. This would have led to the territorial diffusion of cultural norms which then became dominant. This contrasts with an alternative mode of development, termed by Smith 'lateral' amalgamation: in this instance there is a merging of ethnies in which inter-ethnie differences become submerged into a new common identity. In this case the nation could be said to come into being through a process of bureaucratic incorporation whereby localities on the periphery are incorporated into a broader 'national' polity. In essence, however, nations are seen to form around ethnic cores. The reconstruction of these gives rise to the nation. For Smith, nations are not natural entities but they are political–territorial phenomena with a long history.

Modernist theories

Modernist theorists, such as Ernest Gellner and Eric Hobsbawm, argue that nations are relatively new phenomena, 'invented' as recently as the eighteenth and nineteenth centuries although various 'proto-nations' might be recognized prior to that period. The argument here is that nations served, and continue to serve, a particular purpose. They provided a mechanism whereby modernizing capitalist societies could be ordered. Hence, these theories can be termed instrumentalist or functionalist. Rather than being viewed as having deep-seated historical origins, the nation is seen as a construct devised to serve specific purposes. Modernist theorists argue that nations and nationalism merely reflect the needs of a particular economic configuration. In this way they arose as a response to the functional needs of a system rather than the historic inevitability of communal feelings.

During different historical epochs, people owed loyalty to god, to a monarch or to an overlord. In the age of nationalism people owe their loyalty to the collective nation. Dynasty and deity have been replaced by nation. People used to fight for their religion (clearly some still do). Now people have died and continue to die for their country, for their nation's right to self-determination or for its right to defend its territory. Industrialization is seen as the catalyst in this change. It is argued that nationalism only arose in industrialized societies, not in agrarian ones. Gellner (1983) argues that the new social divisions of labour associated with industrial societies needed culturally homogenous literate populations. Nationalism can thus be seen as the result of this objective need. High cultures came to require political support from the masses. In order to acquire this, culture needed a political infrastructure. This developed as nationality, leading to feelings of national identity which served the needs of the industrial age.

An important contribution to the modernist approach on the spread of nationalism has been provided by Benedict Anderson. He suggests that nations can be seen as 'imagined communities' (1991). They are imagined primarily in the sense that they provide feelings of belonging, solidarity and commonality among people who have never met and, in most cases, never will. This shared feeling has arisen through the harmonization of vernacular languages, itself facilitated by the spread of printing and publishing – what Anderson refers to as 'print capitalism'. Allied to the rise of literacy, there was the gradual recognition of a collectivity who shared a language. As Anderson describes it, there were people who could read a book knowing that many others elsewhere were doing the same thing. These people could thus be said to be part of an 'imagined community'. Of course, just as vernacular literature united people it also excluded others.

It follows from this that nationalism is, in the words of Gellner, 'neither universal and necessary nor contingent and accidental' (1997: 10). Certainly the word 'nationalism' does not appear to have entered into common usage until the nineteenth century. The impact of the French revolution is seen by many historians as the catalyst in bringing about a territorially based concept of nationhood. In any event, unlike extreme primordialist ideas, nations and nationalism are seen as political products, not natural characteristics.

The debate between the primordialist–perennialist school and the modernist school, as mediated through Smith and Gellner, is in part a debate over emphasis. From the modernist perspective, nationalism is a force which can utilize pre-existing cultures or, in many instances, completely obliterate them. Thus:

> nationalism is not the awakening and assertion of these mythical, supposedly natural given units. It is, on the contrary, the crystallization of new units, suitable for the conditions now prevailing, though admittedly using as their raw material the cultural, historical and other inheritances from the pre-nationalist world. (Gellner 1983: 49)

In stressing the 'newness' of nationalism, Smith (1986) feels that modernists underplay the role of ethnic origins and the importance of nationalist myths associated with those origins. Modernists do not necessarily dismiss these but they see them as masking more 'real' social forces, rather than viewing them as 'independent' entities.

The construction of a national past is clearly important in maintaining a sense of nationhood. This construction results from the efforts of nationalist ideologues keen to forge a heroic vision of their nation. Instrumentalists tend to view such ideologues who promulgate visions of past national glory as catalysts in present political–national debates utilizing myths of the past to forge a different future. A primordialist or perennialist view is that, while intellectuals clearly do this, they can only do so by virtue of the strength of the myths in the first place. In other words, rather than 'inventing' a national past, the essence of the nation allows the intellectual propaganda to flourish.

It should be clear from the above that there is considerable debate over what precisely a nation is and considerable argument over its origins. Equally, of course, it should be pointed out that the existence of a nation and an associated national identity does not necessarily imply the existence of nationalism. What cannot be disputed, however, is the contemporary importance of nationality in influencing the way in which we view territory and our sense of identification with that territory.

National identity

As has been indicated, the nation refers to a group of people who share particular historical–cultural characteristics or imagine themselves to do so. Nationality refers to the condition of belonging to a nation. At its most basic, nationality can be seen as a mechanism of social classification. People know who they are and who others are (Verdery 1996). We are accustomed to seeing a person's national identity inscribed adjectivally. Thus, David Ginola is a *French* footballer, Youssou N'Dour is a *Senegalese* singer, Carlos Fuentes is a *Mexican* writer. There are two components of national identity, according to Verdery (1996). The first is a collective identity which refers to national characteristics and so-called national traits and may include such things as language and style of dress (see Box 4.1). This is an identity which is shared by the members of the national community. The second meaning to national identity is the individual member's sense of self as a national. An individual's feeling and self-identification as 'English', 'French', 'Spanish' is an important component in their self-perception. It refers to a feeling of belonging to a nation.

In many instances people's national identity may be officially defined in terms of where they were born. However, people may often define themselves in different terms. Many second-generation Irish people living in Britain may see themselves as Irish. Throughout eastern Europe there are numerous nationalities resident within the borders of other states. Thus, there are many ethnic Russians living in the Baltic states of Estonia, Latvia and Lithuania and in the other former member states of the USSR, referred to in Chapter 3. Many ethnic Germans have 'returned' to Germany from Russia following the

---

Box 4.1 Language and national identity

Language is often deemed to be the primary binding mechanism of the nation. Thus, for many Welsh nationalists the preservation and re-enervation of the Welsh language is seen as an integral component in the assertion of Welshness. Kirby (1988), in reference to Ireland, argues that the recovery of the language is an essential element in recapturing national self-esteem, in seeing Ireland at the centre of its own world rather than on the fringes of an English-speaking one.

However, even language, often held up as the primordial glue binding together the nation, is usually itself the result of the creation of the nation rather than its precursor or basic building bloc. Harmonization of vernacular languages, or prioritizing of one form over others, is usually part of the nation-building process, as Anderson (1991) has argued. MacLaughlin suggests that 'cultural characteristics such as national languages, national religious belief systems and national administrative networks were as much the product as the cause of 19th century European nationalism' (1986: 324)

While language is often seen as serving to bind the nation, it can occasionally be counterproductive and give rise to internal tensions. The Welsh language is hailed by some Welsh nationalists as an integral part of Welsh identity

(continued)

(continued)

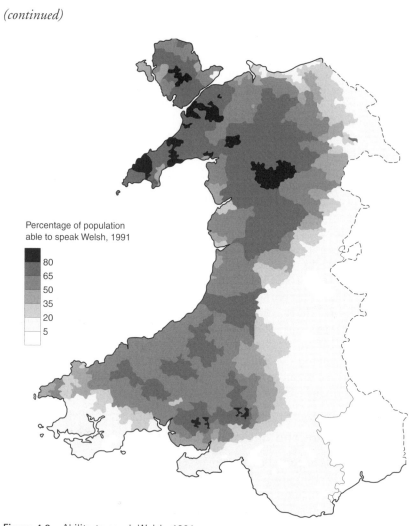

Figure 4.3 Ability to speak Welsh, 1991.
Source: Aitchison and Carter, 1994

(Figure 4.3). However, many Welsh people view the language issue as exclusionary. Throughout much of industrial south Wales many people feel no affinity with the Welsh speakers of the predominantly rural areas of north and mid-Wales. They feel they are viewed as being less Welsh as a consequence of their inability to speak the Welsh language. Hobsbawm (1992) has commented on the emphasis in Wales on ascribing Welsh place names. In cases this becomes slightly absurd, given that some places thus re-named were never Welsh speaking in the first place. Rather than a return to a previous 'national' name, the place is being given a spurious sense of Welshness through the introduction of a new name in order to conform to the project of 'Welshification'.

collapse of communism. They are returning to a country in which they have never lived. It follows that national identity is not simply a function of where a person is born.

Guibernau (1996) sees national identity as composed of five key elements:

1. psychological: consciousness of forming a community
2. cultural: sharing a common culture
3. territorial: attachment to a clearly demarcated territory
4. historical: possessing a common past
5. political: claiming the right to rule itself

Obviously these five characteristics are closely interlinked. Within this milieu elements such as language, religion and social mores may take on particular significance. Many nations are seen to possess their own language, while in some the majority of members adhere to a particular religion. In these cases language or religion may be the key defining characteristic of the nation.

As suggested above, the idea of national consciousness is of major significance. In order to constitute a nation, people must have a sense of themselves as part of that nation. As Muir (1997) points out, national identity is important precisely because people feel it to be important. While it can be argued that there is no objective criteria by which national identity can be measured, its subjective component – the extent to which people believe in it – is very important indeed. A nation is more a mental construct than a concrete reality.

People may also have more than one national identity. Many Indian people in Britain may feel a sense of both Britishness (their country of birth) and Indianness (the country of their parents' birth). This sense of dual national identity has been exemplified recently by the inclusion of a number of English-born players of West Indian parentage in the Jamaican team competing in soccer's 1998 World Cup finals. These included one player who had previously been included in an England squad. This situation even allowed for John Barnes, a veteran of many England internationals but born in Jamaica, to speak on television of his own sense of pride in Jamaica's achievements. A similar strategy had been previously employed to good effect by the Republic of Ireland soccer team in selecting British-born players with Irish parents or grandparents. Ideas of hybrid identities have been common in the United States and are likely to persist, even to grow, in an era of improved communications, increased migration and globalization.

Some people argue that people should choose one national identity above others. This has led to accusations that some black British people have a lessened sense of affinity with Britain, an accusation made explicit in relation to English international cricketers of West Indian origin in an article in *Wisden Cricket Monthly* in 1995 (see Marqusee 1995). The implication is that they are not really British and that they should not be allowed to possess a dual identity. Others, like the writer Salman Rushdie (1992) – himself Indian by birth but resident in the UK – have argued for a tolerance, indeed an encouragement, of dual or even multiple national identities.

Obviously, national identity is by no means the only identity to which people lay claim. Nevertheless, it is an important one in a set of overlapping identities. Thus, people may be male, female, heterosexual, homosexual, black, white, etc. Alongside these various identities people also have a national identity. Which of our many identities predominates will depend on the context. An individual's sense of national identity may be heightened at particular moments. For example, during times of international crisis, such as in the United Kingdom or in Argentina during the Malvinas/Falklands war of 1983, people's sense of national identity may become more prominent. In a less serious context, sporting events, or even occasions such as the much-derided Eurovision Song Contest, may spark heightened feelings of national identity, manifested through support for the representative(s) of the nation.

It is not uncommon for feelings of national identity, otherwise dormant, to come to the fore when visiting (or working or living in) another country. This may, in part, be a consequence of the 'native' population's sense of the 'visitor' as 'different' and in part attributable to the visitor's sense of this difference. Feelings of missing the home country may of course be linked to a sense of missing the comfort and reassurance of familiar surroundings, in terms of familiar faces and places, and may be linked, therefore, to a person's sense of place, as discussed in Chapter 2.

Given the somewhat ephemeral nature of national identity, it is sometimes argued that it is easier to define it in terms of who one is not as opposed to who one is. Being Welsh may be most easily articulated in terms of not being English; to be Portuguese may be conveyed in terms of not being Spanish. National or ethnic identity can be seen as relational. This relational position reflects wider ideas surrounding identity being viewed in relation to an 'other' or 'others' who are seen as possessing a different identity (see Box 4.2). As Edward Said (1995) has argued in his analysis of orientalism, the 'other' becomes objectified. It then follows that our identity is defined in terms of difference from the objectified other. During the colonial era 'Englishness' and 'Frenchness' could be seen in terms of supposedly 'civilized' traits not possessed by those 'others' being colonized. This process allows the members of one nation to view themselves as superior to those of another, thus legitimizing anything from casual disdain, through discrimination, to genocide.

It is also important to note that national identity is shaped, in part at least, by that nation's sense of its own role in the world. Thus, the perception of United States identity is heavily influenced by the perceived importance of the United States in international relations. Many may feel proud of their American identity because of what they see as the important role of the USA in world politics. Others may feel ashamed of their American identity because of aspects of United States foreign policy, for example its record in places such as south-east Asia, central America and the Middle East. By the same token, many British people take pride in their country by virtue of what they see as its former greatness. Britain's imperial past and its economic pre-eminence during the era of industrialization, when it was seen to lead the world, may be a source of great national pride to many. For others, however, this same

Box 4.2   Thai national identity

Winichakul (1996) raises the question of what constitutes 'Thainess' in the construction of Thai nationality. What distinguishes Thais from Burmese, Cambodians or Laotians? In part the answer lies in the contact between indigenous elites and Europeans, in this case the British. In pre-nineteenth-century Siam (the previous name for Thailand) the concept of rigidly demarcated borders was unknown and the idea of overlapping sovereignty was generally accepted and understood within 'frontier' zones. However, Siam found itself in increasing contact with European powers. Specifically, France controlled territory to the east in what is present-day Cambodia, Laos and Vietnam, while the British controlled Burma to the west (Figure 4.4). Relations between the Europeans and those in power in Siam gradually required a firming up of territorial boundaries which had previously been somewhat ephemeral. In order for these boundaries to have meaning on the ground for local people, and to thereby further the 'construction' of the state, a sense of nationhood

Figure 4.4   Thailand.

(continued)

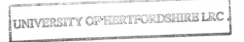

(continued)

needed to be created. From the late nineteenth century onwards there was a need to move towards inculcating a sense of national identity.

In defining 'Thainess' something called 'un-Thainess' also needed to be defined. The former would thus be the opposite of the latter and would embody the positive traits absent in the latter. As Winichakul demonstrates, Thai national identity arises as a relational notion, an identity defined in terms of what it is not as much as what it is:

> The domain of Thainess is also defined by what is 'non Thai'. Once the un-Thainess can be identified, its opposite – Thainess – is apparent. According to a Thai historical perspective, for example, the Burmese were aggressive, expansionist and bellicose, while the Khmer were rather cowardly but opportunistic. The Thai, unsurprisingly, are taken to be the mirror image of these traits – a peaceful, non-aggressive but brave and freedom-loving people, precisely the description in the Thai national anthem. The existence of un-Thainess is as necessary as the positive definition of Thainess. (Winichakul 1996: 219)

This highlights the idea of defining nationality in negative terms; in seeing it quite explicitly in relational terms juxtaposed to other 'inferior' national identities. It also, of course, demonstrates the artificiality of national distinctions for 'what has been believed to be a nation's essence, a justifiable identity, could suddenly turn out to be fabrication' (Winichakul 1996: 220).

---

history may give rise to a sense of shame over acts committed in the name of the British people and the subjugation of the indigenous inhabitants of former colonies. In this way, geopolitics and geopolitical relationships interact with national identity (Dijkink 1996).

Despite the obvious problems which exist in precisely defining national identity that has not prevented attempts to do so. One of the more significant was that undertaken by Stalin who, prior to becoming the leader of the Soviet Union, was its Commissar of Nationalities. Stalin identified four 'scientific' criteria associated with nationality (Alter 1989). These can be seen in terms of four types of community:

- community of language
- community of territory
- community of economic life
- community of culture

The parcelling of Soviet republics around these four criteria by Lenin and Stalin was seen at the time as providing a resolution of nationalist tensions, thereby furthering the communist project. However successful that may have appeared at the time, it clearly has a number of consequences today as ethnic–nationalist tensions pervade many parts of the former Soviet Union, most notably the dispute between Armenia and Azerbaijan over Nagorno-Karabakh, an Armenian enclave located within the borders of Azerbaijan (Figure 4.5).

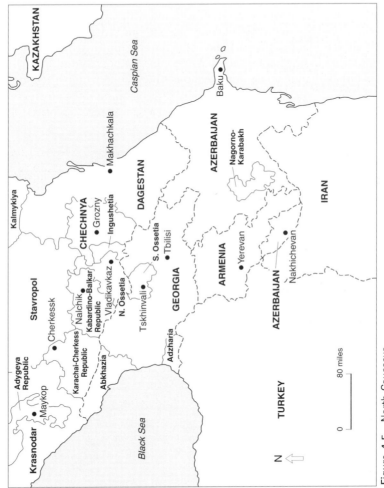

Figure 4.5 North Caucasus.

This serves to demonstrate both the difficulty and the illogicality of defining nationality in precise terms and also the self-defeating and ultimately
counterproductive nature of the exercise in the first place. As Graham Smith
(1995b) has pointed out, the creation of Uzbek, Kazakh, Kirghiz, Tadjik and
Turkmen 'republics' was a Stalinist theoretical construct, not a reflection of
primordial national groupings. Nevertheless, once constructed these territorial
entities subsequently engendered a sense of nationality amongst the populace.

   In summary it can be said that national identity is important to many people
in providing them both with a sense of 'self' and with a sense of being part of a
larger collective. This identity is, of course, a 'felt' one rather than an 'objective'
one. It is also an identity that may be defined oppositionally in terms of what
one is not rather than what one is. Many people may possess multiple national
identities rather than a single unambiguous one. While national identity can
be seen, in some sense, as defying 'rational' categorization, the construction of
national categories creates and reinforces a very real sense of identity.

Nationality and ethnicity

It might be concluded from the above that a national group is similar to an
ethnic group. Da Silva (1996) and others point to the extreme confusion which
exists surrounding the precise meaning of either term and the tendency to use
them interchangeably. Ethnic group refers to a collectivity of people bound
together by elements of culture and by the self-perception that they form a
distinct grouping. On the face of it this sounds identical to a national group.
However, as Miller (1995) and others have pointed out, the two are not
coterminous. Many ethnic groupings exist comfortably within the boundaries
of a larger nation. They do not necessarily wish to form a separate state. While
they may recognize their cultural distinctiveness, they have no expressed desire
to translate this into full political separation from those around them. This is
particularly the case in multi-ethnic societies such as the United States where
numerous ethnic groups peacefully coexist, by and large, without laying claim
to their own national territory. (As will be seen in Chapter 8 they may form
associations with territory at a smaller scale, but this is a different issue.)

   Ethnic groups, in laying stress on shared cultural characteristics and a
common origin, tend to be more exclusive than nations which can incorporate
many different ethnic groups (Kellas 1991). (This is not to say that all ethnic
groups are treated equally within a particular country.) Thus, ethnic groups
tend not to make formalized territorial claims but are chiefly involved with
protecting members' rights within the nation in which they find themselves.
Referring back to Guibernau's five criteria of national identity, ethnic groups
may be said to possess the psychological, cultural and historical elements but
may not necessarily possess the strict territorial element (although there may
be a territory with which they have a distinct linkage through loss or acquisition)
with the consequence that they do not have an overt political–territorial aim.

   Notwithstanding these distinctions, ethnic groups may develop nationalist
aspirations and may seek their own nation. In this way ethnic groups might be

viewed as potential nations, some of whom will one day form nations. Given the historically constructed nature of national identity, it follows that any nation will contain within it the seeds of other nations. In this way separatist movements arise, such as the Basque separatists in Spain, the Albanians of Kosovo in Serbia, Tamils in Sri Lanka, etc. Even newly formed states such as Azerbaijan find themselves faced with secessionist claims, as indicated above. If Quebec attained independence from Canada, it is quite probable that it would be confronted with claims for Iroquois independence.

National identity could thus be seen as a more overtly politicized version of ethnic identity. The mechanisms which function so as to bring about the transition from ethnic identity to national identity are certainly not clear. As suggested above, territory – more specifically, the desire or claim to politically control a certain territory – is an essential component of national identity but not of ethnic identity. Thus, nationality is an intrinsically geographical phenomenon, more so than is ethnicity.

Nationalism

As suggested earlier, many people identify strongly with the nation to which they feel they belong. At its most extreme, this may be apparent through a willingness to fight or even die for one's country. In more peaceful circumstances this sense of identification may be manifested through such activities as support for national sporting teams or individuals and through singing, or observing solemnity during, the playing of the national anthem. Even at the comparatively innocuous level of self-definition as 'Scottish', 'French', 'Vietnamese', etc., people almost unthinkingly see themselves in terms of their nationality. This sense of identification is seen to reflect an ideology of nationalism. It is an ideology in the sense that it encapsulates a set of beliefs and practices which people come to accept as 'natural'.

The communal sense of identity that is at the heart of nationalist ideology gives rise to a sense of a 'national will' which unifies all members of the nation. This reification of the nation presents it as a 'natural' and largely unchanging phenomenon. As will be seen, this is very far from being the case. In this way, international football supporters talk of 'their' victories, 'their' defeats, etc. For some supporters, defeat for their international team is a defeat for them. It is they who have lost, not merely the players on the pitch. Similarly, a desire to attack another nation can be put into practice through an assault on its people.

It is important to observe that, as implied earlier, the nation refers to a social collectivity rather than territory. Nevertheless, it is a territorial concept in that the group of people concerned feel an attachment to a particular territory and national disputes invariably centre on struggles over the control of land. Because of this, nationalism is a territorial ideology reflecting attachment to a particular space and one which, in its more 'active' form, seeks to maintain or to attain political independence (and in some cases dominance) for the nation and, hence, for its territory. Anthony D. Smith defines nationalism as 'an ideological movement for attaining and maintaining autonomy, unity and identity on behalf of a population deemed by some of its members to constitute an actual or potential "nation" ' (1991: 73). This reminds us that a sense of nationhood, while it may assume an 'essential' quality, is ultimately a tool useful in the attainment of particular political goals.

Types of nationalism

There are many different classifications of nationalism proposed by various authors. One division quite often used revolves around a distinction between a civic–territorial nationalism and an ethnic–genealogical one. This is utilized by Anthony D. Smith and it suggests that in many countries, most notably within western Europe, national identity is derived on the basis of where one is born. Thus, if a person is born in Britain they are rendered British. Of course a person may also acquire British or US nationality through the fulfil-ment of criteria based on behaviour (residency for a certain period of time, language proficiency, etc.). This is seen to contrast with many eastern European countries and with Germany where national identity is based on ancestry. Attempts to resolve conflict in the countries of the former Yugoslavia centre around complex questions of national or ethnic identity which are not easily resolvable through reference to place of birth. In Germany, citizenship is available to those of German origin even though their families may have lived elsewhere for centuries. This has led to an influx of ethnic Germans from Russia and Poland; people accorded German citizenship yet who have never lived in the country. By way of contrast, ethnic Turks who have lived for many years in Germany, or have even been born there, could not until recently obtain German nationality.

This distinction between an ethnic (eastern) and a civic (western) nationalism carries certain connotations. In Kohn's (1967) and Plamenatz's (1976) terms the ethnic variant is, perhaps understandably, seen as less progressive than the civic version. The latter is seen as furthering the cause of democracy while, in contrast, the other is seen as 'negative', being based, it is argued, on emotion. As a result ethnic nationalism is deemed irrational, unlike the civic-based nationalisms of the west, of which Britain, France and the United States are seen as the exemplars. As a consequence of this, there is a tendency to frown upon the apparently tribal-like nationalisms of the Balkans and elsewhere in eastern Europe. This is in contrast to the supposedly fully evolved civic nationalism of western Europe and North America. However, many western countries might do well to reflect on the existence of ethnic tensions and overt racial discrimination within their own borders. While this has not necessarily led to ethnic cleansing and mass murder, it does, nevertheless, suggest that the civic–territorial nationalisms have been less than fully successful at integrating all members of society into a common and egalitarian polity. Certainly, many people of West Indian or Asian origin in the United Kingdom and many Algerians and Moroccans in France might take a somewhat more jaundiced view. It can also be argued that civic nationalisms themselves are built around particular 'ethnic myths'. For example, notions of British tolerance and the 'stiff upper lip' are often presented as 'superior' traits to the apparent ethnic or 'tribal' myths which are seen to generate conflict. This ignores the constructed nature of these supposedly national traits (O'Dowd 1996).

Another basic distinction within the literature on nationalism is that between forms of 'top-down' nationalism on the one hand and 'bottom-up' nationalism on the other. The former occurs where a political elite attempts to ensure a nationalism which inculcates loyalty to the state. In these examples the creation of the state pre-dated a sense of national consciousness. In this way, nationalism as an ideology follows state formation rather than being a precondition for it. England and France are cited as examples of this. In these instances what has occurred is 'the promotion of ethnic images and affinities and their application to the state' (Slowe 1990: 91). Tom Nairn (1977) sees this as national determinism or bourgeois nationalism. In order for this to occur it requires the invention of a nationalist ideology which serves to bind the nation. Nairn argues that this is what most western European nationalisms are about. Certain key traits are identified and are promulgated in such a way as to serve to bind the nation together. In this way the inculcation of national identity serves a clear political end, namely the maintenance of state hegemony.

Alternatively, we can identify national self-determination or nationalism from below. This is a nationalism born of a desire to shake off a particular imposed rule. It is a nationalism emanating from the 'periphery', quite often applied to Third World countries, but with European examples such as Basque nationalism and that of the 'Celtic fringe'. This is a nationalism generally constructed in opposition to rule by a colonial power. The post-colonial states of Africa, Asia and Latin America provide many examples of this form. Colonial boundaries cut across pre-existing tribal boundaries but nations were created out of these artificial collectivities in order to rally support to the anti-imperialist cause. In this case nationalism was an ideology created to accord with the 'artificial' boundaries of the imposed state. Nevertheless, once independence was achieved, the newly engendered national consciousness could then be used to bind the residents of the state together. Boundaries do not conform to any meaningful 'national' divisions. This means the end product of 'bottom-up' or 'peripheral' nationalism is exactly the same as 'top-down' forms, namely the inculcation of a national consciousness in order to bind the state together. What has happened in much of Africa, Asia and Latin America appears to have been the adoption of particular 'modular forms' derived from the European experience (Anderson 1991; Davidson 1992). In other words, post-colonial states have adopted a territorial model of the nation.

Nationalism as progressive or regressive?

Because of its association with war and brutal animosity between people, nationalism is quite often seen in negative terms. Both world wars have been seen to result from nationalist territorial claims. More recently nationalism has been seen in highly negative terms in the light of events following the collapse of communism in eastern Europe. Tensions began to flare up in parts of the former Soviet Union with a rise of ethno-nationalist feelings. More prominently, the bloody unrest in the former Yugoslavia concerning territorial disputes involving Serbia, Bosnia-Herzegovina, Croatia and Kosovo has tended to lead many to condemn nationalism as a very negative human phenomenon. With on-going conflict concerning Northern Ireland, this picture of nationalism as a negative and disruptive phenomenon is reinforced. The utilization of national identity in order to justify racism and various forms of racial supremacy, whether in the era of formal colonialism or in the present-day rhetoric and actions of far-right political parties and movements, adds to the uncomfortable feelings attaching to national identity.

It is undoubtedly true that nationalism can give rise to extremely bloody consequences. The pursuit of the nationalist ideal can lead to such practices as ethnic cleansing. In the past it served as a useful adjunct to bolster the colonization of supposedly 'inferior races'. Calls for the repatriation of 'foreigners', such as those by the British MP Enoch Powell in the 1960s and 1970s, and more recent versions of this in the rhetoric of the far right in Britain, France, Austria and other countries, represent attempts to 'purify' the territory of the nation, ridding it of 'external' impurities. There is a danger attaching to nationalism of it being either perceived as, or in reality being, an exclusivist ideology in which people who do not belong to the nation are excluded from the benefits of membership and are denied equal status. This leads to racism, ethnic tensions and, in its most virulent forms, programmes of ethnic cleansing, expulsions and repatriations. As a consequence, nationalism itself is seen as a source of evil.

In a similar vein, Chatterjee (1995) has argued that nationalism is seen to be something of an unwanted virus spreading from the periphery and beginning to infect the core. It is seen as an ailment of eastern Europe and the periphery of western Europe (Celtic fringe, Basque country). It is also significant, as Seton-Watson (1977) and others have noted, that patriotism which can be interpreted as a form of state or 'official' nationalism (Kellas 1991) is seen as 'good' while nationalism is 'bad'. Thus, defence of an existing set of national institutions can be seen as patriotic but attempts at instilling a sense of nationhood among people with no existing political institutions tend to be seen in a negative light. Clearly this is because events likely to lead to political instability are generally seen (at least from the perspective of those in power) as unwanted, while those which reinforce the status quo are generally viewed in a positive light.

Related to this, Billig (1995) argues that there is a strong tendency to see nationalism only in its 'hot' forms, i.e. when it leads to violent conflict. As a consequence, nationalism is seen as a 'problem' afflicting the former Yugoslavia, Northern Ireland and the Basque country. However, it is not seen as a phenomenon in relation to England or Spain. As Blaut (1987) has pointed out, there is a tendency to focus on 'small' nationalisms rather than 'big' ones. Nationalism is perceived as a force driving secessionist movements and leading to instability but it is not seen as a force for stability. The fact that present-day states are underpinned by a sense of nationhood, subtly instilled through various mechanisms, is ignored. This tendency to see nationalism as a phenomenon of the periphery is bound up with imperialism and conquest. As Smith puts it, 'those whose identities are rarely questioned and who have never known exile or subjugation of land and culture, have little need to trace their "roots" in order to establish a unique and recognizable identity' (1986: 2). Within the United Kingdom, nationalism tends to be seen as something which is peculiar to the Scots, Welsh and Irish while curiously not afflicting the English. They are quite happy to see British identity prosper, because, to a considerable extent, it is seen as identical to their own English identity (see Taylor 1991). Indeed, as Eric Hobsbawm has noted, the very term 'English nationalism' 'sounds odd to many ears' (1992: 11). Despite this, ideas of national chauvinism and jingoistic behaviour continue to exist. Current alterations in the nature of the UK, with devolution for Scotland and Wales, seem likely to promote considerable debate on the nature of Englishness and Britishness.

As Anthony D. Smith (1996) points out, most historians appear to view nationalism in negative terms. Eric Hobsbawm is a case in point. To him and others, nationalism appears to be an intrinsically 'bad thing'. In part, it might be argued, this reflects the quest for academic objectivity and detachment. It might also be argued that it reflects certain academics' disdain for what is seen as vulgar, sentimental and, perhaps, irrational behaviour. In addition, nationalism is often associated with the political right and, as already suggested, can carry with it exclusionary ideas and can act as an excuse for racist and xenophobic behaviour. While Hobsbawm's laments over the bloody effects of nationalism appear to reflect a commendable vision of an egalitarian world shorn of national chauvinism, they also play down the understandable appeal of nationalist rhetoric to people whose rights and freedoms have been trampled on.

Related to this is the presumption that extreme nationalism is aberrant or irrational behaviour (Kellas 1991). Some academics, most notably Conor Cruise O'Brien, heavily influenced by events in his native Ireland, have suggested that nationalism is an entirely irrational and regressive phenomenon. Ironically, as Ernest Gellner (1994) has pointed out, O'Brien's stance is self contradictory. If nationalism is as deeply rooted as O'Brien suggests, it can hardly be seen as aberrant behaviour. For advocates of the 'nationalism as aberrant' school of thought, there appears to be a view that something else ('real socialism' or 'rational liberalism' perhaps) should have happened. The idea of nationalism as a form of aberrant behaviour is akin to the view that nationalism is essentially a disease, the implication being that the disease needs to be eradicated (Miller 1995). It might be argued that extreme nationalism (however that might be defined) may reflect a perfectly rational set of motives – defence of home, job, security, power, etc. Those who have pursued Irish unity through military means may well perceive (rightly or wrongly) continued British rule in Ireland as being against their better interest. Thus, there may well be more than a supposedly irrational attachment to territory or a misty-eyed romanticism underlying nationalist movements (although these are undoubtedly quite often present and indeed are often invoked as part of the 'nationalizing' strategy). For Tom Nairn (1997) a world of an increasing number of nations is not 'unreal' or aberrant. Rather, it points to the powerful meanings which attach to national identity, meanings which arise for a variety of reasons. While there are many negative phenomena associated with nationalism (racism, xenophobia) Nairn asserts that the nation is far from dead and it may offer hope to some, thus suggesting that a total demonizing of the phenomenon is not apt.

It is possible to see national differences in a more positive light. A distinction can be made between an exclusive or narrow conception of nationality or a more inclusive or broader conception. The former revolves around racism and xenophobia while the latter is a pluralistic version. This can be related to another distinction: that between a nationalism which asserts cultural distinctiveness but which does not involve an assertion of national superiority and a nationalism which is chauvinistic. The acceptance of national differences in terms of culture, language, etc. can be seen to contribute to international diversity and cultural enrichment. This contrasts with national chauvinism which asserts that one nationalism is superior to all others. Manent (1997) argues that the nation can be the basis for a set of positive values and a framework for meaningful democratic institutions, arguments akin to those of Miller (1995). Rather than being automatically associated with regression, nationalism might be linked to progressive democratization.

Leaving aside the more overt manifestations of national identity, the phenomenon is reinforced in ways which seem quite ordinary and mundane and appear, on the surface at least, to have nothing to do with politics (see Box 4.3). Many people feel 'English', 'Irish', 'Spanish' and feel they are different in many respects from members of other nations. This does not necessarily lead to conflict but merely to a recognition of differences. Conflicts only arise in situations where territorial expansion into another's 'space' occurs or where national chauvinism exists where members of a national group perceive themselves to be superior in some way to members of one or more other national groups. None of this is to ignore the fact that people may identify with their locality, county, region or more than one nation, as suggested earlier. There are levels of territorial identity and the intensity of feeling will vary depending on the specific context. Regardless of the academic 'opinion' of it, nationalism is a phenomenon requiring serious study. As Hugh Seton-Watson has argued, 'merely to inveigh against nationalism does little to help the human race'

Box 4.3   Banal nationalism

We are aware of the more overt ways in which nationality is asserted: through calls to arms in times of crisis and through the singing of national anthems, etc. However, as Michael Billig (1995) has argued, nationalism is an ever-present phenomenon and the nation is re-produced in many less obvious everyday ways. The flying of flags on public buildings and many other mundane everyday occurrences may well go unnoticed and unremarked (Plate 4.1). However, through their very ordinariness, they inculcate a sense of national

(continued)

(continued)

Plate 4.1   United States' flags flying, Wall Street, New York.

identity. This is what Billig refers to as 'banal nationalism'. He argues that in western states nationalism, far from being a thing of the past, is in fact an endemic condition. The nation is constantly being flagged, thereby rendering it easier to mobilize the citizenry in its support during times of crisis.

In a myriad little ways people are silently complicit in the reproduction of their nation. These unremarked aspects of nationalism, these many and varied means through which the nation is constantly reaffirmed, serve as useful rehearsals for those moments when more overt forms of national mobilization are needed. Through the banal affirmations an unquestioning support can more readily be attained. This adds considerably to the relevance of Ernest Renan's (1882, cited in Billig, 1995) view of the nation as a daily plebiscite whereby the willingness of the people to believe in its importance served to ensure its existence.

(1977: xii). It might be appropriate to view nationalism itself as a neutral phenomenon, one which can be put to use for what might be seen as either progressive or conservative ends.

Functions of nationalism

It should be obvious from the foregoing discussion that irrespective of the view that is held on the origin of nations, and regardless of whether it is viewed as a progressive or regressive phenomenon, nationalism serves a number of purposes in modern societies. A first set of functions relates to the nation being a convenient mobilizing tool to help the state to achieve its objectives. In this way it is useful in maintaining hegemony and legitimacy for the state. An alternative set of functions is more subversive in that nationalism can be used as a mechanism for righting other wrongs. Thus, greater democracy or equality for marginalized groups may be fought for under the banner of nationalism.

It is clear that nationalism is a very useful mobilizing force, particularly if it can be shown that 'the nation' is under threat. It has already been suggested that nationalism as an ideology has served to perpetuate certain forms of territorial control. In the contemporary world, Saddam Hussein can utilize a version of nationalist rhetoric in urging the people of Iraq to support him in his defence of the nation against the external threat posed by the United States. Similarly, United States regimes can refer to a defence of American interests as a mechanism for mobilizing public opinion over acts of aggression towards other countries. In reality, what may be happening is a defence of US business interests.

The distinctions between government, state and nation are sometimes deliberately blurred by governments as a mechanism for undermining opposition. Governments may occasionally claim that opposition to them is unpatriotic. In this way, opposition to United States foreign policy is sometimes portrayed as un-American. To oppose its foreign policy (a criticism of government) is sometimes used by the administration as an indication of a lack of patriotism (support for the country) in an attempt to undermine the validity of the arguments being used. However, the majority of such criticism, whether from the left or the right, is not about undermining the nation, but rather about aspects of state policy or the nature of the state. It is not necessarily opposition to the United States as a political entity (although it might be). Indeed, right-wing opposition within the USA is often of a 'patriotic' nature arguing against state intrusion but supporting 'American values'.

National identity can be used to maintain political hegemony. A political elite can invoke nationalist rhetoric in order to maintain its own hegemonic control within its territory. Nations can be constructed around particular sets of ideas. Thus, the Nigerian writer Wole Soyinka describes a nation as 'a gambling space for the opportunism and adventurism of power' (1996: 286). He argues that the geographic spaces that constitute present-day African states are spaces in which people are marginalized and exploited while those doing the exploiting invoke the rhetoric of nationalism to mask their otherwise naked display of power. In the United States to be labelled as anti-American is a useful method of minimizing protest against unpopular policies. If opposition to specific measures or a specific regime can be presented as opposition to the nation then the arguments may appear less popular. The nation, as a concept, is, thus, a very useful means of political control. It enables those in power to get ordinary people to identify with the nation and specific measures can be justified on the nation's account. Thus, people are encouraged and expected to 'rally round the flag'. In the famous words of John F. Kennedy: 'ask not what your country can do for you, ask what you can do for your country'.

Nationalism can be sold as a sop to those who feel alienated. In this way the poor and the unemployed can take pride in their nationality. People are urged to look on 'higher' things rather than the mundane events of everyday life. People may be elevated above the tedium and hardships of the present, thereby rendering them more subservient to those who rule in the name of the nation. The nation is, thus, an agent of legitimation and mobilization, a means by which powerful elites can retain or gain power. Working-class Protestants in Northern Ireland can feel proud of their religion and their British heritage. In this way they can be mobilized as a political force in defence of what they see as their right to remain within the UK. As Soyinka (1996) argues, clearly with his native Nigeria very much in mind, political elites can invoke nationalist rhetoric and play on otherwise meaningless differences in a strategy of divide and rule. As a result, much of Africa remains divided against itself, thereby impeding action which might otherwise lead to some form of more progressive politics. The divide and rule strategy can be used in cases to create overt strife. This can be used to justify harsh measures to restore stability. Hegemony can be retained through the invention of instability.

In a different way nationalism can be seen as a useful tool in attempting to redress what are seen as social injustices. In many instances what appear to be disputes centring on national or ethnic tensions may be more to do with the marginalization of particular groups and may be connected with social exclusion as much as with nationalism *per se*. Tamil separatism in Sri Lanka may owe its strength more to the marginalization of Tamils than to any innate sense of wishing for full political autonomy. The latter may, in effect, be a device for redressing the former. Kurdish calls for an independent Kurdistan reflect not just a territorial wish to have their own political space but also reflect the social, economic, political and cultural marginalization of Kurdish people in the countries in which they find themselves, most notably Turkey and Iraq. Viewed in this way, nationalism can be seen as a territorial strategy useful in achieving social and economic objectives.

Returning to Wole Soyinka, he stresses that his Nigerianness is one that he accepts as a 'duty' by virtue of being born into that geographic space. This means that he can work with others also born into that space; working with them, not towards exclusivist ends, but rather towards extending opportunities and widening access to resources; not just for those within the bounded space that is contemporary Nigeria, but also outwards to embrace the rest of Africa. In other words, Soyinka deems it useful to use the framework of Nigeria for what he sees as worthy objectives. Like the Nigerian regime, he sees the nation in functional terms, but, unlike them, he is seeking very different ends. For the government, 'Nigerianness' is a mechanism by which they can control, while for Soyinka it is a framework within which one can work towards liberation.

Finally, as the above examples may suggest, it might be argued that nationalism is in some ways a necessary precondition for democracy. If there is a sense that people should have a say in how they are ruled, this implies some idea of who the 'people' actually are. Despite the view commonly held that nationalism and democracy are mutually antagonistic, writers such as Nodia argue that some sense of nationalism, based on a collectivity of people within a defined territory, underpins all forms or attempts at democracy: 'nationalism is the historical force that has provided the political units for democratic government' (1994: 7).

Summary

This chapter has provided an introduction to the concept of the nation and the associated territorial ideology of nationalism. Theories on the origins of nations and the role and purpose of nationalism have been reviewed. Regardless of the debates surrounding the origins of nations and the variety of views surrounding the phenomenon of nationalism, it is clear that nations do exist. People see themselves as belonging to a nation. Whether this is viewed as rational, irrational, positive, negative, progressive or regressive it remains an indisputable phenomenon. It is equally clear that nations are 'produced' rather than being 'natural'. It follows that a national identity is a constructed one. Consequently, nationalism serves particular functions at specific times. Despite the recognition of nationalism as a historically contingent phenomenon which serves particular ends, 'none of this makes the nation 'unreal' for an ordinary man [*sic*] born into a concrete society, culture, and state, and faced with concrete choices on the social and political as well as the spiritual and existential planes. The nation need not be "rational" in order to be "real" ' (Nodia 1994: 11). Equally indisputable are the powerful connections between national identity and territory. These links between place and nation are explored more fully in the next chapter.

Further reading

There is a wealth of literature dealing with nationalism. Particularly important are the works of Anthony D. Smith, especially *The Ethnic Origins of Nations* (Blackwell, Oxford, 1986) and *National Identity* (Penguin, London, 1991), and of Ernest Gellner, in particular *Nations and Nationalism* (Blackwell, Oxford, 1983), *Encounters with Nationalism* (Blackwell, Oxford, 1994) and *Nationalism* (Weidenfeld and Nicolson, London, 1997). Eric Hobsbawm's work is also influential; see in particular *Nations and Nationalism since 1780. Programme, Myth, Reality* (2nd edition, Cambridge University Press, Cambridge, 1992). A good review of key debates is presented in Smith's

Nationalism and Modernism. A Critical Survey of Recent Theories of Nations and Nationalism (Routledge, London, 1998). See also the work of David Miller, especially *On Nationality* (Clarendon Press, Oxford, 1995) and Hugh Seton-Watson, *Nations and States. An Enquiry into the Origins of States and the Politics of Nationalism* (Methuen, London, 1977). Benedict Anderson's *Imagined Communities. Reflections on the Origin and Spread of Nationalism* (Verso, London, 1991) is an important contribution to the debate on the origins of nations. Montserrat Guibernau offers fresh insights in *Nationalism. The Nation-State and Nationalism in the Twentieth Century* (Polity Press, Cambridge, 1996). See also Tom Nairn, *Faces of Nationalism. Janus Revisited* (Verso, London, 1997) and Michael Billig's important *Banal Nationalism* (Sage, London, 1995). The edited volume by Gopal Balakrishnan, *Mapping the Nation* (Verso, London, 1996) brings together a range of historical and contemporary perspectives.

On nationalism and ethnicity see Walker Connor, 'A nation is a nation, is a state is an ethnic group is a . . .', *Ethnic and Racial Studies* 1978, 1(4): 377–400, J.G. Kellas, *The Politics of Nationalism and Ethnicity* (Macmillan, Basingstoke, 1991) and K.M. da Silva's chapter 'Ethnicity and nationalism' in *Between Development and Destruction. An Enquiry into the Causes of Conflict in Post-Colonial States* (Netherlands Ministry of Foreign Affairs, The Hague, 1996) pp. 109–25, edited by L.van de Goor, K. Rupesinghe and P. Sciarone. The last of these deals with issues of 'Third World' nationalism as does Basil Davidson in his numerous writings such as *The Black Man's Burden. Africa and The Curse of the Nation-State* (James Currey, Oxford, 1992) and Partha Chatterjee, *The Nation and its Fragments. Colonial and Post-colonial Histories* (Cambridge University Press, Cambridge, 1993).

Links between geopolitics and nationalism are the subject matter of Gertran Dijkink's *National Identity and Geopolitical Visions. Maps of Pride and Pain* (Routledge, London, 1996). For a discussion of Welsh nationalism see Piers Gruffudd, 'Remaking Wales: nation-building and the geographical imagination, 1925–50', *Political Geography* 1995, 14(3): 219–39. On Thai nationalism see the article by Thongchai Winichakul, 'Siam mapped. The making of Thai nationhood', *Ecologist* 1996, 26(5): 215–21.

Chapter 5

Nationalism and the importance of place

The power of the nation to induce strong emotional and physical reactions varies from country to country and between individuals within the same 'nation'. The sight of football players and fans proudly singing their national anthem while waving or saluting 'their' flag, perhaps with tears in their eyes, is a powerful reminder of the pervasiveness of national identity. This pervasiveness reflects the success of efforts to create a strong national consciousness. This sense of the nation is constantly constructed and re-constructed in ways sometimes overt and sometimes banal, as suggested in the previous chapter. In order to do so nations require features which can be utilized in the process of affirming or re-affirming nationhood. There is a need for a national past which is seen to provide the glue which holds the nation together. In tandem with this national past is a national geography built around particular places and utilizing explicit territorial allusions.

People, events and places can be put into the service of nation-building and affirmation. This means that nations require a history built around these elements in order to sustain their existence and meaning in the eyes of their nationals. Places are, by their nature, geographical entities while people live in places and events occur in, or are associated with, places. This implies that a 'national' history works in conjunction with a 'national' geography to present a vision of the nation. Thus, while history is a key element in the construction of the nation, geography is also important. In this way particular sites or particular landscapes – spaces and places – become imbued with meaning:

> a certain tradition of images, cults, customs, rites and artefacts, as well as certain events, heroes, landscapes and values, come to form a distinctive repository of ethnic culture, to be drawn upon selectively by successive generations of the community. (Smith 1991: 38)

Within this mix, diverse elements such as battles, language and landscape come together to produce the nation and are utilized in order to preserve its territorial integrity. This chapter explores the use of people, events and places in the making and re-making of the nation. First, the importance of a 'national' history is assessed. Subsequently, the connections between particular places and the nation are explored. The chapter continues by considering the significance of territory in the construction of the nation in four rather different contexts: the Balkans, Ireland, England and Israel–Palestine. The key focus in each of these examples is the importance of territorial imagery in the construction and reproduction of the nation and/or the utilization of territorial strategies in order to reinforce or to resist a particular political configuration. Territory, it is argued, is important both symbolically and in practical terms.

The importance of history

A version of the nation's past needs to be brought into being. All nations require a past to justify their current existence and to provide a rationale for territorial claims. Fact, folklore and fiction combine to produce and reproduce a sense of nationhood; myths and legends are an important part of nation-building. 'National' histories tend to present a relatively seamless narrative through which the members of the nation can trace their collective past. This is not the same as saying that an 'accurate' version of history is important. Rather a 'suitable' past is required and 'if there is no suitable past, it can always be invented' (Hobsbawm 1998: 6). Further, given the role and importance of myths in nation-building, 'inaccurate' histories are perhaps crucial. As Ernest Renan observed towards the end of the nineteenth century, 'getting its history wrong is part of being a nation' (1882, quoted in Hobsbawm 1992: 12). In fact, it may be **necessary** to get it wrong. From a functionalist perspective these elements of the past, whether real or invented, are utilized for contemporary purposes:

> To demonstrate that tradition is wrong or invented does not put an end to this story. A claim to national independence does not fall simply because its legitimising version of national history is partly or wholly untrue – as it often is. The sense of belonging to a distinct cultural tradition . . . can be subjectively real to the point at which it becomes an objective socio-political fact, no matter what fibs are used for its decoration. (Ascherson 1996: 274)

As noted earlier, particular people may be invoked in the propagation of the nationalist myth. Certain key historical figures are assumed to embody the nation – Nelson in England, Lincoln in the USA, William Wallace in Scotland (Plate 5.1), Owyn Glyndwr in Wales. These figures are mythologized, or practically invented in some instances, and become symbols of the nation's past when they fought for its freedom or performed glorious deeds on its behalf. It is not just obvious political figures who can be taken as symbolizing the nation. Even contemporary personalities from beyond the world of politics or the military can be invoked as symbolizing the nation. Sportspeople are often seen as carrying the mantle of the nation and their actions may be interpreted as symbolizing the national spirit. The athlete Sonia O'Sullivan, draped in the Irish flag as she does a lap of honour around an athletics stadium following a race victory, is seen as a victorious vision of Ireland. The French football team pictured clasping the 1998 World Cup were seen as the embodiment of a victorious (and multiracial) nation.

Just as certain individuals are taken as emblematic of the nation, particular events may be seen as vital to the national project. Scottish people are assumed

Plate 5.1   Memorial to William Wallace, Stirling, Scotland.

to identify with the Battle of Culloden while the Highland Clearances are seen as an event which exemplifies the suppression of Scotland by England. The victory of William of Orange (a Dutchman) over King James at the Battle of the Boyne in 1690 (well before the partitioning of the island) is seen as a defining moment in Irish history underpinning the right of the unionist people of Northern Ireland to remain under British rule. Sites of major battles, as well as buildings associated with key individuals, become national monuments – places seen as being of vital historical importance and, hence, of crucial significance to the nation's present. In a world ever more fully in thrall to heritage in its various guises, the preservation or construction of a national heritage built around people and events is of ever-increasing significance (Lowenthal 1998).

It follows from the above that the veracity of the role of individuals or the nature of key events is not what is important. Rather, it is the mythological interpretation which is placed upon them. Events and people become 'traditionalized' in order to celebrate the nation. In this way it can be said that nations are constructed through the invention of tradition (Hobsbawm and Ranger 1992). Particular customs or events are portrayed as stretching back into antiquity to a time immemorial, to the primordial nation. Examples of this invention include the Scottish kilt, Welsh Eisteddfodd and the British coronation ceremony, all of which are nineteenth-century inventions (although the last two are, in part, based on earlier formats). All of this implies a construction of the nation out of various historical and geographical fragments. In the flowery prose of Anthony D. Smith, nationalism provides 'cognitive maps and historical moralities for present generations, drawn from the poetic spaces and golden ages of the communal past' (1991: 69). Once again this highlights the emphasis on the invocation of both the geographic and the historical in the reproduction of the nation.

Territory and nation

Within the discourse of the construction of the nation, it is obvious that territory is of huge importance. This is more than the simple fact of territory being a fundamental requirement for the nation's existence. Historical fact and myth concerning particular places are key elements in the national imagination. The importance of key individuals and events in nation-building has already been noted, individuals and events usually associated with particular places. In addition to these indirect place associations, there are two main ways in which territory features in nationalist narratives. First, there are numerous references to the 'generic' territory of the nation. Allusions to the national soil abound within such discourses. The second territorial element is the importance attaching to specified places. For example, past Serbian occupation of present-day Kosovo serves as a useful justification for territorial control over an area which is almost exclusively occupied by ethnic Albanians. The connections between people and place, discussed in Chapter 2, suggest that a sense of place, when linked to the political project of nationalism, results in the sustaining of evocative images of place which come to symbolize the nation.

Landscape and territorial imagery are very important elements within national identity. Certain places may acquire huge symbolic significance. Particular parts of the national territory may acquire a significance as the presumed 'zone of origin' of the nation; its original heartland. Thus, the Canadian north with its vast, remote, rocky and forbidding image is seen as symbolizing an independent Canada. The 'taming' of the American west means that not just the 'pioneers' heading westwards but also the landscapes through which they travelled assumed significance in the nation-building project. In versions of Welsh nationalist discourse the mountains are seen as the heart of the nation, somehow symbolizing a Wales untainted by outside, specifically English, influences. Thus, Plaid Cymru chose as its original symbol an idealized representation of mountains. A leading figure in the party during its early years suggested 'the mountains! The perpetual witnesses of our history, and the unchanging background of our language' (cited in Gruffudd 1995: 224). Another nationalist spoke of the mountains thus: 'To us, the mountain is the bread of life, and is a holy sacrament. Our lives are woven into its essence' (cited in Gruffudd 1995: 228). As Gruffudd concludes, 'remote areas away from the anglicizing influences of the accessible lowlands harboured the "national character" and nationalist politics, therefore, centred on the defence of that cultural integrity' (1995: 236). Indeed, in some discourses, the land itself assumes the role of a sentient being, it has memories embedded within it: 'its memories are sacred, its rivers are full of memories, its lakes recall distant oaths and battles' (Williams and Smith 1983: 509).

Specific places, rather than generic features, also come to assume symbolic importance. In this way, the White Cliffs of Dover are taken as one of the symbols of England and, by extension, Britain. They may be seen as the embodiment of the nation in a place, for example, the Wailing Wall in Jerusalem, at least among Jews; the Dome of the Rock may be seen in an equivalent light among Palestinians. In this way national identity is often defined in terms of the territory historically inhabited by the nation. Particular places may be utilized in this celebration of the nation. In Eritrea the currency is named after a place – Nakfa – which was the site of an important battle during its secessionist war with Ethiopia in the 1980s. Thus, place impacts even on national currency.

The importance of territory and of specific places is emphasized in many national anthems. These pieces of music, the most overt means by which the nation is symbolized, quite often contain territorial references. These may take the form of generic allusions to soil and land or more specific references to particular places or landscape features such as mountains or rivers. In this way, songs seen as the nation's musical signature often have a strong territorial base which evokes images seen as part of the essence of the nation. Hobsbawm has written of the attempt to inculcate a sense of Austrianness through the use of territorial imagery in the first short-lived national anthem of post First World War Austria (a remnant of the former Habsburg Empire). In Hobsbawm's words, this anthem involved 'a travelogue or geography lesson following the alpine streams down from glaciers to the Danube valley and Vienna' (1992: 92). The current national anthem of Austria provides an example of the use of generic landscape features as symbols of the country:

> Land of mountains, land of rivers,
> Land of tillage, land of churches,
> Land of iron, land of promise,
> Motherland of valiant sons.

Norway's anthem too refers to its citizens' love of this country:

> Yes we love with fond devotion
> This our land that rises rugged, storm-scarred,
> O'er the ocean with her thousand homes.

In a similar vein 'Flower of Scotland' evokes a rugged rural landscape, one which is defended against the marauding external foe (England):

> O Flower of Scotland
> When will we see your like again?
> That fought and died for
> Your wee bit hill and glen
> And stood against him
> Proud Edward's army
> And sent him homeward
> Tae think again.

For sheer elevation of territory to a mythical status, Chile's anthem takes some beating

> Chile, your sky is pure blue
> So sweet as the breezes that roam
> Over your fields, embroidered with flowers
> That angels might make thee their home.

Such hyperbole serves to manufacture an almost magical landscape. Similar sentiments abound in Denmark:

> There is a lovely land
> Whose charming woods of beeches
> Grow near the Baltic strand, grow near the Baltic strand.
> It waves from valley up to hill
> Its name is olden Denmark.

While the above refer to landscape features in quite general terms, many anthems refer to specific places seen as integral to, or in some way the embodiment of, the nation:

> Proudly rise the Balkan peaks
> At their feet the blue Danube flows
> Over Thrace the sun is shining
> Pirin looms in purple glow.

In this verse, Bulgaria lays claim to its defining territory with mountain and river signifying the nation. Of particular significance here is the reference to Thrace, an ancient territory now largely in Greece. In this way Bulgaria lays claim, emotionally at least, to this place now lying within the borders of another country. Similarly the anthem of Croatia refers to rivers:

> Sava, Drava, keep on flowing,
> Danube, do not lose your vigour,
> Deep blue sea, go tell the whole world,
> That a Croat loves his homeland.

All three rivers referred to serve, at some stage, as Croatia's borders. The Drava, Danube and Sava form part of the country's borders with Hungary, Serbia and Bosnia-Herzegovina respectively. Throughout many national songs the defence of the homeland is a recurring theme; the salvation, or maintaining, of territorial integrity is seen as crucial rather than the attaining of democracy or a particular form of government. In this way the Mexican anthem addresses the territory directly:

But should a foreign enemy
Dare to profane your soil with his tread
Know, beloved fatherland, that heaven gave you
A soldier in each of your sons.

These examples also demonstrate the elevation of quite ordinary landscapes into something else. There is nothing intrinsically unique about mountain ranges, coastlines or woods, but in the nationalist discourse these features assume emblematic status. In highlighting the importance of particular places, it is illuminating to present examples of the way in which geography and a sense of place become integral to the creation and sustaining of a national imagination. The four examples which follow are derived from somewhat different political contexts. Serb identity is one which currently attempts to justify a geographically larger stake for itself in the rubble of the former Yugoslavia; English national identity is complicated by a confusion between English and British while Irish identity is constructed in the light of a colonial struggle against Britain. Finally, there is a brief exploration of the significance of territory in the case of Israel–Palestine where two national groups lay claim to the same national space.

Territory and nationalism in the Balkans

Following the collapse of the communist regimes in much of eastern Europe in the late 1980s, the USSR began to disintegrate, resulting in the creation or re-creation of the Russian Federation and fourteen republics. In addition, territorial changes also took place elsewhere in eastern Europe. The Czech Republic and Slovakia emerged out of what had been Czechoslovakia. More painfully, the formerly non-Warsaw-Pact state of Yugoslavia began to fall apart with its constituent 'nations' endeavouring to go their own ways. The result was the eventual creation and reluctant recognition of Slovenia, Croatia, Bosnia-Herzegovina and Macedonia, with Serbia and Montenegro remaining to form what is now a much territorially diminished Federal Republic of Yugoslavia.

The state of Yugoslavia was itself created at the end of the First World War out of territory which had long been contested between the Ottoman and Austro-Hungarian empires. It is a region containing a complex variety of ethnic groups which have a history of both antagonism and peaceful intermingling and intermarriage (Figure 5.1). However, the disintegration of Yugoslavia led to an immediate rush by non-Serbs to secede from the Belgrade-centred state. The nature of Balkan history means that territorial divisions between distinct groupings are not easy to identify, with many competing claims over particular places.

Unlike the 'velvet divorce' of Czechoslovakia, the disintegration of Yugoslavia was far from peaceful and was highly contested. Outright war involving Bosnia-Herzegovina, Serbia and Croatia resulted in hundreds of thousands of deaths, ruined villages and infrastructural breakdown. It is estimated that the Bosnian conflict has resulted in 300,000 deaths (Steele 1999). It also saw the entrance into common English language usage of the term 'ethnic cleansing', as entire towns and villages were subjected to mass extinctions of the 'enemy' ethnic grouping. The resolution of the Bosnian conflict, as laid down in the Dayton

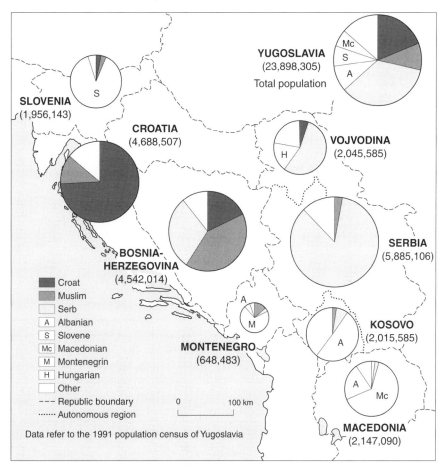

Figure 5.1 Ethnic composition of former Yugoslavia.
Source: Popovski, 1995

agreement of 1995, contains within it the seeds of its own possible destruction. The state of Bosnia-Herzegovina consists of two distinct entities; a Muslim–Croat Federation and a Bosnian Serb Republic (Republika Srpska). This renders the current state of Bosnia a somewhat unstable mix with competing Muslim, Croatian and Serb sectors (Figure 5.2). In 1999 NATO conducted a war against Yugoslavia, ostensibly to prevent the massacres of ethnic Albanians in Serbia's southern province of Kosovo. Prior to the hostilities, the area's population was 90 per cent ethnic Albanian and, following the withdrawal of autonomy in 1989, many objected to Serb rule.

While conflict has raged over enclaves of Serbs and Croatians within Bosnia and the designation of particular zones as 'belonging' to one or another ethnic grouping, there is another vitally important territorial component to the conflict which extends beyond the present day ethnic composition of the population. Central to these disputes have been claims and counterclaims to places seen as historically belonging to one group or another. Underlying this ethnic tension,

Figure 5.2   Bosnia-Herzegovina.
Source: Glenny, 1996

Serb 'geography' is of prime importance. White (1996) has highlighted the significance to Serb nationalists of places which are not actually in present-day Serbia. Thus, Serb folk songs make mention of places in neighbouring states of the former Yugoslavia – Croatia, Bosnia-Herzegovina – and even in Greece and Albania. Songs refer to battle sites, to mountains and to rivers in territory outside Serbia's current borders. White quotes Stevan Kacanski's 1885 verses

> Round the Struma and the Vardar
> a lovely flower blooms,
> it is the flower of the Serbian tsar:
> the holy blossom of Tsar Dusan.
> Round the Struma and the Vardar
> bloom the flowers of the Serbian tsar.

In this extract two rivers are used to evoke an image of a Serb heartland; but the Struma and Vardar run through the territory of present-day Macedonia. The point is that, although the subdividing of the former Yugoslavia along broad ethnic lines which resulted from the Dayton agreement might be seen to reflect current ethno-national distributions of population, it ignores the fact that certain places are imbued with meaning for Serbs, even though few Serbs may actually live in these places. A concentration on the ethnic composition of a particular locality fails to take account of the symbolic meaning of particular places (White 1996).

Of particular significance in this regard is Kosovo which, under the former Yugoslavia, had been an autonomous region within present-day Serbia. Currently, it is a sight of on-going ethnic and political tension as the vast majority of its inhabitants are ethnic Albanians, not Serbs. Serbian attempts to remove its autonomy following the collapse of the federal Yugoslavia have been met with political and military opposition, with the Kosovo Liberation Army claiming to defend the ethnic Albanians against the colonising Serbs. Ultimately, this led to outside intervention with NATO launching air attacks on Serbia and the deployment of a peace-keeping force on the ground. In Serb mythology their defeat in a battle by the Turks at the Battle of Kosovo Polje in 1389 is seen as a formative element in their national identity. To concede control of the location of their heroic struggle, even though few Serbs live there, would be seen as sacrificing an integral part of Serb territory and, through it, Serb identity. It would be construed as a betrayal of their own history. The symbolic significance attaching to this place has proved a useful device through which the Serbian leader Slobodan Milosevic can sustain political support built around nationalist–territorial rhetoric.

Similarly, Macedonia was the centre of a medieval Serbian Empire. Macedonia is particularly interesting because to Serbs, Bulgarians and Greeks it is deemed historically to be an integral part of their respective nations. In fact, Greece has objected to the use of the name Macedonia as the title of the newly independent state: hence its official designation (in English) as the Former Yugoslav Republic of Macedonia, as noted earlier. The importance of territory as a site of key events in nation-building is augmented by the fact that many of the nation's heroes have come from Macedonia. Thus, Alexander the Great, a Greek hero, came from what is present-day Macedonia, as did the Serb hero Prince Marko: 'To say that Macedonia is not rightfully part of Serbia is to say that Prince Marko was not really Serbian' (White 1996: 1). Further confusing the issue here is that the Bulgarians also claim Prince Marko as a national hero.

In summary, the ethnic conflict in the former Yugoslavia is a clear example of the symbolic importance of territory and it demonstrates the manner in which territorial arguments can be utilized in order to gain, or to maintain, political dominance.

English nationalism and territorial imagery

In a speech in 1993, then British Prime Minister John Major spoke of those things which, in his view, exemplified Britain. It was a 'country of long shadows on county grounds, warm beer, invincible green suburbs, dog lovers and pools fillers' (cited in Billig 1995: 102). He spoke in a way which implied that all Britons would agree with him. This evocation serves to illustrate the commonplace aspects of national identity, often associated with everyday spaces. Clearly the view is a very partial one, evoking rusticity and presenting a certain notion of timeless landscapes as symbols of the nation (to say nothing of the reflection of a male, rather than female, sense of affinity).

In any analysis of national imagery it is important to observe what is included (or excluded). Different strands can be identified in the territorial imagery normally associated with England. The first is a confusion between Englishness and Britishness. Major's vision of Britain highlights a central issue in that the exemplars used to illustrate Britishness might be seen as peculiarly English. Billig also cites Tony Blair, then in opposition, outlining his vision of Britain as 'nation of tolerance, innovation and creativity . . . [with] . . . a great history and culture' (1995: 105). This can be further seen as a subsuming of British identity into an English one. Considerable confusion exists in distinguishing English culture from British culture and in determining where English national identity ends and British national identity begins. For many people in England there is a tendency to assume that English identity is synonymous with British identity. For people in Scotland and Wales the distinction is clearer. Generally, they will see themselves as either Welsh/Scottish or British, or both. In the latter case which of the identities, British or Welsh/Scottish, takes precedence will vary. However, for many English people, the distinction is unclear. Many of the symbols of Britain and Britishness are in fact emblematic of England – cricket, stiff upper lip, warm beer, village greens, rustic villages, gentle landscapes. The use of this imagery is designed to foster a sense of British national identity. But it is significant that in so doing there is the imposition of one of the subsumed identities, namely Englishness, as the identifier for the wider sense of Britishness. Englishness simply 'is'; it is the norm. To many this merely reinforces a sense of English cultural imperialism over Wales and Scotland. The territorial images of Britishness are those which might be seen as quintessentially English.

The preoccupation with such things as English sporting success (most notably in football) appears to predominate over a subordinate concern with the 'minor' nations within the kingdom. It might be argued that this reflects population levels and is compensated for by regional coverage on television, etc. However, this does not get beyond the fact that, at a 'national' level, England is the nation, in a way in which Scotland and Wales most definitely are not. This often gives rise to a certain bewilderment. English football fans cannot understand (or more likely understand but do not care) why their team is resented by many Welsh and Scottish people (and by many others as well). In part at least these attitudes would appear to reflect a colonial past which continues to permeate the present.

Leaving aside the confusion between Englishness and Britishness, another contradiction emerges. English nation-building, like other national histories, is heavily dependent on antiquity. Somewhat perversely, many images of England, as with other countries, are rural and 'old' as distinct from urban and modern. In part, this is bound up with a response to the processes of industrialization and urbanization. Constructions of the urban as 'evil' and contaminated have been mirrored by the creation of the 'rural idyll' whereby all that is 'pure' and 'natural' is seen to be associated with the rural landscape and with rural life. These wholesome images are then taken as embodying true Englishness. England is not a rural country yet many of its popular images are rural. In this way the paintings of Constable are taken as representing Englishness and there is a long history of the use of idyllic landscape imagery to conjure up the nation (Plate 5.2). Ironically, as Taylor puts it, 'the pioneer of industrialization and the most urbanized country in the world is idealized in rural terms' (1991: 151). This linkage between idyllic rural images and national identity is by no means unique to England (see Box 5.1).

Plate 5.2 *Hereford Dynedor and the Malvern Hills from Haywood Lodge*, painting by George Robert Lewis, 1815; Tate Gallery, London 2000.

Box 5.1 Arcadian images

In many countries there is an image of rural areas as the 'real' country; the one which existed prior to the contaminating effects of urbanization. Linked to this is the elevation of agriculture as a wholesome and rewarding way of life. These images and beliefs have their roots in Arcadian notions of rural innocence, of people uncontaminated by external urban influences. In much of Europe, the second half of the nineteenth century was characterized by rapid industrialization and urban growth. Cities were seen by many as smelly, dirty and 'unnatural' and this led to an anti-urban backlash. In Britain and elsewhere, versions of this rural idyll exist, although they are tempered in instances by the recognition of the existence of rural poverty and out-migration from rural areas. In Wales, Plaid Cymru originally encouraged moves 'back to the land' and gave agriculture an exalted status. The mountainous north and middle of the country were seen as the heartland of the Welsh nation; zones relatively uncontaminated by outside influences. In post-independence Ireland there was an explicit invocation of the rural as the repository of something truly Irish. These rural images also serve to suggest a 'natural' state or a state of innocence. Within a nationalist discourse these allegedly uncontaminated places symbolize the nation in its original and 'pure' form.

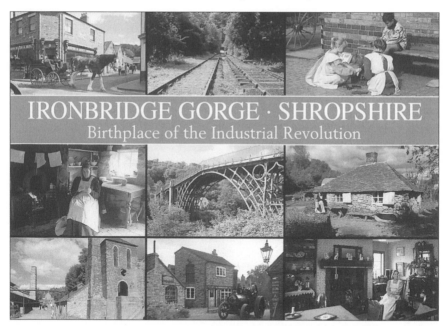

IRONBRIDGE GORGE · SHROPSHIRE
Birthplace of the Industrial Revolution

Plate 5.3   Postcard of Ironbridge, Shropshire, England. © J. Salmon Ltd., Sevenoaks, England.

Even when industrialization is being considered particular places can be utilized in presenting a vision of the past. In this way, Ironbridge in Shropshire has become a 'museum' of the industrial revolution with a number of different sites illustrating various aspects of England's status as the 'first industrial nation' (Plate 5.3). In this celebration of industrialization, this one place has become a symbol for a wider vision of England's industrial heritage.

There are many sanitized versions of the nation's past which serve to generate feelings of warmth, nostalgia and goodness. A feeling is inculcated of how lucky people are to be English, to be proud members of a proud nation. Allied to symbols of monarchy, stately homes, alleged virtues of tolerance and 'fair play', territorial images and national characteristics create a view of England–Britain which is seen as comforting and reassuring and, strangely, one which is seen as non-nationalistic. With the recent elections to a Scottish parliament and Welsh assembly (see Chapter 7), questions of English identity seem likely to incur closer scrutiny in the near future.

Ireland

Territory and territorial imagery have played a prominent role in the creation of Irish identity while territorial strategies play a part in the on-going conflict in the north of Ireland. Traditionally Ireland has portrayed itself (and been portrayed by others) as a rural place. This has often been used to promote an image of tranquillity and quaintness; a sort of 'other-worldliness' in which Ireland is seen as something of a refuge from the modern world; a place in

Plate 5.4   Near Cahirciveen, County Kerry, Ireland.

Plate 5.5   *In the West of Ireland* painting by Paul Henry.

Plate 5.6   Deserted house, west of Ireland.

which traditional values of family, co-operation and friendliness abound. In the 1930s the then Taoiseach (Prime Minister) Éamon de Valera enunciated his vision of Ireland as a place of 'comely maidens dancing at the cross-roads'. This image of rusticity and youthful innocence (coupled with a devout Catholicism) was a metaphor for the nation. In part at least, this emphasis on rural imagery is associated with an opposition to an urban colonizer, England. To a considerable extent, urban areas in Ireland tended to be seen as those places most 'contaminated' by foreign influence. Thus, a true or 'real' Ireland was seen as a rural one, and a western one, the part of the country seen to be least influenced by the English (Plate 5.4).

Following from this, rural Ireland has been presented as embodying the essence of the nation. The west of Ireland, with its rugged mountains, is seen as the 'real' Ireland, a place with a certain spiritual mystique which has been the inspiration for literature and other art forms. Artists such as Paul Henry have used the west coast and off-shore islands as the foci of their work (Plate 5.5). The west is seen to embody Irishness; mystical, romantic but also remote, desolate, de-populated. All of these are images which themselves are seen to capture Ireland's existence at the edge of Europe and also its apparent sadness as a migrant society emptied of its people (Plate 5.6).

Plate 5.7 Irish postcard.

These romanticized visions are those carried with them in their heads by those who migrate. They are also the predominant images of the island used in tourist brochures, on postcards, etc. Such images tend to play down the extent to which Ireland has become a much more urban society and one which, in an era of globalization, is ever more fully integrated into a wider world (Plate 5.7).

Ireland's territorial integrity has obviously been disrupted by the north–south partition of the island in 1921. This is frequently presented in terms of an 'unnatural' division of a small island. Northern Ireland comprises six of the nine counties of Ulster, the northernmost of Ireland's four historic provinces (Figure 5.3). In story and song, continued British occupation of the north is seen in terms of an unwarranted disruption to the island's territorial integrity. The four provinces are often referred to in songs as the 'four green fields' with the on-going territorial dispute over the north seen in terms of recapturing the fourth green field (see Box 5.2).

Box 5.2   Political conflict in Northern Ireland

The conflict over Northern Ireland is a territorial dispute over whether the region should remain under British jurisdiction or should become part of a united Ireland. All of Ireland for a time was under British rule, having become fully incorporated into the United Kingdom of Great Britain and Ireland in 1800. Following a succession of rebellions, a settlement of 1921 saw the creation of an independent Ireland consisting of twenty-six of its thirty-two counties (known since 1948 as the Republic of Ireland). The remaining six counties (Northern Ireland) became a semi-autonomous part of the United Kingdom. This region was ruled by its own parliament in Belfast controlled by unionists (so called because they wish to remain part of the United Kingdom). The unionist government consistently abused the democratic rights of the minority Irish nationalist population (who wished to see a united Ireland), leading to periodic bouts of civil unrest. These became prominent in the late 1960s, culminating in sectarian rioting and the deployment of the British Army. Ultimately, the British government suspended the Stormont parliament and ruled directly from London, a situation persisting until the present.

The present stage of the conflict can be dated to 1969 when civil rights protests were met with a violent response by the police and this led to a resurgence of support for republican groupings, most notably the Irish Republican Army (an organization which itself has displayed a number of fractures). This is a paramilitary organization dedicated to the ending of British rule in Ireland and the re-unification of the country. This is implacably opposed by unionists, including a number of paramilitary groupings such as the Ulster Volunteer Force (UVF) and the Ulster Defence Association (UDA). While the protests of the late 1960s were more about civil rights than national rights, the conflict soon took on a distinctly nationalist orientation. Since the upsurge in violent conflict in 1969, almost 3,500 deaths have occurred, with many more people seriously injured. Conflict has also been extended to Britain, and to other countries, with attacks by republican groups on British Army bases in Germany, for example.

The opposition between unionists and nationalists is overlaid by a religious divide. Most Unionists are Protestants of English and Scottish ancestry, many of whom are the descendants of people 'planted' in Ireland by the colonial power during the seventeenth century. The majority of nationalists are Catholic. However, these divisions, while very real and resulting in many overtly sectarian attacks and murders, are underpinned by a sense of political injustice on the part of nationalists and by a sense of encirclement on the part of unionists.

In 1998 a peace agreement was signed in Belfast which seeks to create a cross-party Northern Ireland Assembly together with a number of cross-border all-Ireland bodies. The agreement has been opposed by one major unionist faction and by a number of small republican groupings. Progress on operationalizing this agreement has been painfully slow, with unionists continuing to argue for a decommissioning of weapons by paramilitary groups before allowing republicans to take their ministerial positions in a Northern Ireland 'cabinet'.

Figure 5.3 Ireland.
Source: Graham (1997)

Territory continues to play a significant role in the conflict. Even the very act of naming the territorial entity is fraught with difficulty and ambiguity. Republicans tend to prefer the terms 'north of Ireland' or 'Six Counties' to 'Northern Ireland' thereby expressing their non-recognition of its legitimacy. Unionists prefer the term Ulster, partly, it might be surmised, because it does not have the word 'Ireland' in it and tends to confer a sense of being a place separate from the rest of the island. The Battle of the Boyne and the locations of other battles hold important symbolic meaning for loyalists and representations of King William at the Battle of the Boyne feature on many banners carried at loyalist Orange Order parades. This Protestant organization holds marches each summer in towns and villages throughout Northern Ireland. It is seen either as a defender of Protestant religion and culture or as a potent symbol of oppression, depending on one's standpoint. Their annual marches have in many instances led to flashpoints, most notably at Drumcree, near Portadown, over recent years. The right to march in certain areas and down particular streets is presented by the order as a basic right but, when the areas concerned have significant nationalist populations, such marches are often interpreted as expressions of loyalist power. The corollary of this is that republicans are not wanted and are not safe on those same streets. This could be interpreted as a territorial strategy employed to remind nationalists of their 'place'. The sense in which residents in some areas may feel 'hemmed in', or are actually physically prevented from leaving, adds to the direct territorial expression of loyalism. The painting of kerbstones in blue also indicates loyalist support. Individuals may decorate their houses with flags or other symbols such as union jacks reflecting their political views. The painting of wall murals can also be seen in terms of laying claim to space and a depiction of political loyalties (see Chapter 8). In traditional songs, places assume prime importance. The most famous orange song is The Sash, in which particular sites of conflict in preceding centuries are utilized to depict the loyalist motto of 'no surrender':

> It is old but it is beautiful, and its colours they are fine,
> It was worn at Derry, Aughrim, Enniskillen and the Boyne,
> My father wore it as a youth in bygone days of yore,
> And on the Twelfth I love to wear the sash my father wore.

While references to specific places and the utilization of territorial strategies have been central to unionist tactics, it has been argued that a key problem for unionists is their inability to marshall an adequate set of place images of 'Ulster' (Graham 1997). Geographic imagery of Ireland as a particular type of place has been the preserve of republicans. Unionists, in seeing themselves as British, have dissociated themselves from images of Ireland and Irishness, but they equally do not fit neatly into notions of Britishness and cannot readily identify with place images of 'Britain'.

Palestine–Israel

The creation of the Israeli state in 1948 out of the British mandated territory has led to an as yet unresolved conflict over territorial control and national identity. The state of Israel was created in order to provide a homeland for the Jewish diaspora scattered throughout the world. The creation of such a state was given added significance following the extermination of millions of Jews by the Nazis during the Second World War. Although there were other possible locations for a Jewish state Israel was chosen, primarily because of its symbolic and historic connection with the Jewish people. The philosophy of Zionism portrayed it as essential that Jews acquired their historic territory promised them by God. The territory of Israel assumes huge importance for Jews based on their religious history. In biblical terms Israel is the 'promised land' and it contains within it the essential ingredients of the Jewish state. The problem that exists, however, centres on the reality that another recognizable national group, the Palestinians, also inhabit that space (Figure 5.4).

Originally twin states were envisaged but Israel simply claimed the bulk of the territory for itself. That Jews formed only a minority of the population in this former colony of the Ottoman empire in 1948 was an irrelevance. The new state embarked on a territorial strategy designed to cleanse the territory of its non-Jewish inhabitants. Many were simply forced into exile as villages were literally destroyed and replaced by Jewish ones. The land was, in effect, made for the new immigrant Jewish population. Israel was, quite literally,

of writing Russia is again engaged in military conflict in Chechnya, while a war of secession is being fought within neighbouring Dagestan (see Figure 4.5).

These are examples of what Muir (1997) is referring to when he recently re-introduced the notion of 'effective national territory' into political geography. This idea was originally devised by Preston E. James (1959) and is based on the observation that some states may not have full or effective control over parts of their territory. There may be 'holes in the state', as Muir terms them. Examples of this include parts of Colombia deemed to be effectively controlled either by drug barons or by guerrilla armies, parts of Sudan controlled by warlords and, prior to its secession from Moscow, Chechnya, as indicated above. In a number of countries, particularly within the 'Third World', significant parts of the national territory may be beyond government control with militias under the direction of various groupings effectively policing these areas. In 1999, there was considerable media interest in the murder of tourists in southern Uganda by presumed members of the former Rwandan army who appear to operate across the borders of Uganda, Rwanda and the Congo Democratic Republic (formerly Zaire). It appears to be the case that these border zones are not totally under the control of the states to which they officially belong.

Such 'holes in the state' may result from a variety of factors. It follows from earlier discussions dealing with functionalist theories of the state that those states experiencing greater centrifugal forces will be more apt to be under considerable pressure in continuing to exist. Countries with significant ethnic or national divisions, linguistic differentiation or major regional inequalities might be seen as relatively unstable. In this way some parts of the state may become virtually ungovernable by the central authority and may often come under the control of groups not aligned to the state as currently constituted. This section explores how the sometimes overlapping issues of substate nationalism, language and core–periphery contrasts affect the state before a consideration of state responses to these issues.

Substate nationalism

Pressures on the state are most commonly due to secessionist nationalism where national groups wish to see the creation of alternative political units carved out of part of the territory of the existing state. Classic examples of this are the struggle for Basque separatism which seeks a Basque state wholly independent from Spain (see Box 6.1). The Basque conflict has led to numerous military attacks in Madrid and elsewhere and assassinations of Spanish politicians.

Northern Ireland is another classic example of separatist nationalism. Nationalists within the north of Ireland wish to see the region become part of a united Ireland. This irredentist claim is opposed by unionists who see themselves as British and wish to remain so (see Chapter 5). In its attempts to control Northern Ireland, the UK has on occasions effectively conceded as 'no-go' areas parts of Belfast and Derry where Irish nationalist or republican support predominates. Within this context a good example of the lack of

Box 6.1 Basque separatism

The region referred to as the Basque country is part of northern Spain and extends northwards into southern France. It is an area with a distinct culture, including its own language. Historically, the region has enjoyed periods of relative autonomy but it has never had complete political independence and resentment has been continuous as the region's identity has been devalued within the process of Spanish state-building. Calls for a separate Basque state arose in the early part of the twentieth century when Franco's fascist regime opposed any autonomy for the region and was distinctly hostile to Basque culture and language. The Basque language, like Catalan, had been effectively banished as Castillian was the promoted 'national' language of Spain (Laitin, Solé and Kalyvas 1994).

This overt oppression helped bolster support for militant nationalism and saw the rise of Euskadi ta Askatasuna (ETA – Basque Homeland and Liberty), formed in 1957 out of a variety of pre-existing factions, as a military organization aimed at secession from the Spanish state. ETA has been involved in many military attacks in Spain and many have died in the ensuing conflict.

When Franco's reign came to an end in the 1970s, Spain moved towards a federal solution to the problem of regional instability and granted the Basque country a degree of autonomy, within a federated Spanish state. Thus, the Basque country can provide its own services and even has its own police force. Nevertheless, control over things such as foreign policy and general economic planning resides with the central government in Madrid. Despite this, many still see complete independence as the ultimate goal. Militant ETA attacks have resumed following the ending of a ceasefire in the late 1990s. Walls in the towns and cities of the Basque country continue to be emblazoned with graffiti asserting the right to independence (Plate 6.1).

Plate 6.1   Graffiti in support of Basque independence, Bilbao.

Plate 6.2 Cardboard cut-out of AK-47 attached to lamppost in a republican area, Newry, Northern Ireland. Photo by Paul Anthony McErlane.

effective control relates to south Armagh where Irish language road signs have been erected by local people. While an armed British presence is maintained through a fortified base in the village of Crossmaglen, the area is not under effective control. Soldiers are helicoptered in and out because of the threat to them from snipers on the ground. In this and other parts of Northern Ireland, security checks and road blocks have been mounted in the past by paramilitary organizations effectively 'patrolling' such areas and thereby signifying them as their territory (Plate 6.2). On-going punishment beatings being carried out by both loyalist and republican paramilitary groupings reflect a form of policing of their own 'turf'. Clearly, if such lack of total internal control reaches beyond a certain point, then the state's territory may eventually be reduced *de facto* as well as *de jure*. In places such as south Armagh, Irish flags are flown as a symbol of rejection of British rule and in favour of Irish rule. These symbols once again prove important signifiers of territorial desires. These and other examples point up the potentially unstable nature of the state and they serve to highlight the fault lines which exist within individual states.

While the Basque and Irish situations have produced militant conflict for the Spanish and UK states, milder substate nationalisms exist in both countries. In the case of Spain, significant Catalan and Galician minorities exist in the north–east and north–west respectively, while in the UK debates continue over Scottish and Welsh independence. Here, there is a desire among Welsh and Scottish nationalists to achieve independence from the UK. This has resulted in a devolved parliament for Scotland and an assembly for Wales following referenda on the issue in 1997.

Within Africa there are numerous examples of ethnic differences giving rise to secessionist tendencies. Eritrea's protracted war of secession is a prime example. While Eritrea succeeded in its secessionist struggle, failed attempts abound. Examples include the Ibo insurrection within Nigeria leading to the Biafra War of 1967–70. Part of the problem for African states is that they were created by external colonizing powers with the result that the concept of the state and its present configuration may have little effective meaning for its citizens, despite nation-building attempts by political elites. Other forms of loyalty exist and ethnic divisions may render the state inherently unstable. Conflicts within Rwanda, Burundi, Congo Democratic Republic, Nigeria and elsewhere reflect an unease with those states as currently constituted. Within Somalia the situation provides a rich breeding ground for warlords who are happy to use territory as a 'gambling space' for their own desires, to borrow Wole Soyinka's (1996) phrase. The view adopted by many observers is that numerous African states are virtually ungovernable with portions of territory controlled by various armed factions and state institutions effectively ceasing to function. Forrest (1988) distinguishes between 'hard' and 'soft' states. In part the 'soft' states have never had stable mass parties; rather, they have had attempts to construct 'national' parties out of many varied factions and groupings. In some instances these states may eventually 'collapse' (Muir 1997).

While this idea of inherently unstable states is quite useful and appears to reflect political–territorial realities, there are two risks attaching to these views. The first is the danger of assuming that all African states, or indeed all 'Third World' states, are the same. This homogenizing view ignores the differences in history and geography of a complex part of the world. The second risk is the one of assuming the problem is intrinsically 'African'; that it somehow reflects Africans' inability to govern or to look after themselves; that it represents the absence of a viable civic culture. Instead, the fault may well lie elsewhere. As Richard Muir expresses it:

> European avarice created a world of sovereign states, each one cast in the European legal mould. The people of the newly emancipated African colonies could either accept the world of the national sovereign actor or else be denied a place on the diplomatic stage. Not surprisingly, the political codes and institutions which had evolved slowly according to the shifting nuances of European life often failed to flourish when superimposed on African society. How unjust that Europe expects the world to believe that this is all somehow the Africans' fault. (1997: 212)

This can be read as Africa utilizing European modular forms with, in many instances, quite disastrous consequences (Davidson 1992). In addition, it is clear from earlier discussions that unstable states are a political reality in Europe as well as in Africa.

Language divisions

Within the complex mix of substate ethnic and nationalist tensions, cultural differences may assume huge importance. In this case elements such as language may act as a centrifugal force causing instability. Belgium is an example of a country with a significant language divide between a French-speaking Wallonia and a Flemish-speaking Flanders. Since the 1960s the country has followed a strict language equality policy which also includes a degree of recognition for German, spoken in areas along the German border. The recognition of the language divide was cemented with the adoption of a federal structure in 1993 resulting in a tripartite territorial division between Flanders, Wallonia and Brussels which, as the capital city, is officially bilingual although it is located within Wallonia (see Figure 3.3). Flemish is the official language of Flanders, French is the official language of Wallonia. Tensions have arisen at various times which strongly suggest that identification with language zone is, perhaps, stronger than identification with Belgium. Under these circumstances, some feel that there is no sufficiently strong underpinning for the Belgian state. Murphy (1993) has pointed to the limited interaction between residents of the two main regions and has also highlighted the fact that, once regional boundaries are firmed up, as they were in the 1960s in the case of Belgium's linguistic regions, they tend to reduce interaction across those boundaries. Just as national boundaries serve to create difference, so too do internal borders. As suggested in Chapter 3, boundary creation gives rise to its own dynamic whereby previous divisions become institutionalized and enshrined as part of everyday existence. In this way people's everyday lives become territorialized as a consequence of the political–territorial framework within which they live.

In the case of Belgium there is limited transgression of the language borders. However, in Switzerland, a country with four official languages, there is less linguistic tension and language does not appear as a major fault line in Swiss politics. In part at least, this is related to a different internal political territorialization which cross-cuts, rather than mirrors, the language divide and thereby serves to weaken language divisions rather then to highlight them. The example of Belgium seems to suggest that differences may become more deeply inscribed, and subsequently reinforced, through territorial divisions. In this way those divisions which are the reason for federalism in the first place become reified and reproduced as a consequence of federalism (Agnew 1995). A sense of place and identity is reinforced through the federal structure, further highlighting the way in which politics and territorial identity are intermeshed.

In Canada, a language issue also threatens the continuance of the Canadian state. In Quebec, there is a French-speaking majority and Canada has adopted a bilingual policy in order to satisfy the demands of those in Quebec who feel that for far too long Canada was an anglophone state which essentially denigrated 'French Canada' and Quebec's distinctiveness. In the early 1990s a referendum on independence was narrowly defeated, roughly 55 per cent opposing a break-up. Nevertheless, a distinct divide between 'French Canada' and 'English Canada' still pervades political and cultural life (Williams 1995). Indeed the Quebec case is only one of a range of potential secessionist issues facing the Canadian state. In 1999, a self-governing Inuit territory was created in the north.

Core–periphery contrasts

Socio-economic differences within countries have been seen as a consequence of the development of cores and peripheries with some regions remaining marginalized while others are seen to be economically dominant. Some have likened this to a form of 'internal colonialism' (Hechter 1975). These core–periphery tensions can also place pressure on the state with calls for secession emanating from either the periphery or, perhaps less frequently, the core. High levels of disaffection in particular regions might lead to calls for secession or for greater autonomy. It is argued, as we have seen in previous chapters, that peripheral nationalisms may have as much to do with regional inequalities as with any innate sense of nationhood. Peripheral places are dependent on cores and a sense of resentment may develop towards an 'imperial' centre which is seen to exert control over them. Current moves at asserting Cornish identity are strongly related to ideas that the county is being peripheralized with the closure of its last functioning tin mine and the encouragement of tourism, perceived by many as signifying its declining importance. It is argued by some that Welsh economic peripherality has strongly influenced Welsh nationalism just as much as a sense of Welsh identity. Marginalization of Flanders has in the past been a mobilizing force in assertions of Flemish identity, thereby contributing to the territorial divide within Belgium. A different type of periphery dissatisfaction occurs in Spain where the apparent prosperity of the Basque country and Catalonia has, in the past, been seen as a reason for separatism, in order to avoid these riches being siphoned off by the Spanish state.

State responses to secessionist tendencies

In the face of secessionist or divisive pressures which threaten its territorial integrity, there are a variety of responses which a state may employ. We can follow the framework outlined in Chapter 3 when discussing state responses to secessionist tendencies. In brief, the state will devise a variety of mechanisms to resist disintegration. These responses may be either coercive or adaptive. In the former case, internal threats to stability may be dealt with harshly through the pursuit of a military strategy, as in the case of Serbia's handling of unrest in Kosovo. Overt military tactics are usually accompanied by harsh legal measures implemented through a punitive judicial system. The death penalty, lengthy prison sentences and institutionalized torture may all be elements within this oppressive response. The treatment of Kurdish separatist groups and the general suppression of Kurdish culture in Turkey is an example of a coercive reaction to secessionist tendencies. The obvious intention here is to annihilate resistance through the overt suppression, execution or jailing of dissidents and those seen to be fomenting strife and/or to create an atmosphere of fear whereby people are discouraged from taking action against the state.

However, there are more subtle, although perhaps equally effective, responses which fall short of outright coercion. The opinions, beliefs and aspirations of secessionist groups may be discredited or cast in a negative light through various forms of state propaganda or the dissemination of misinformation. In some cases outright censorship of these views may be employed. For a time, legislation in both the UK and Ireland made it illegal for television and radio to broadcast interviews with people who were members of Irish republican organizations. Whether overtly coercive or not, the intention underlying these methods is the same; to preserve the political and territorial integrity of the state. Through various mechanisms, an ideology of the state as the 'natural' unit is promoted and anything which destabilises it is seen as 'unnatural'. This corresponds to Gramsci's ideas of the state retaining its hegemony through more subtle ideological rather than coercive means.

Alternatively regimes may pursue a policy of accommodation or appease-ment of secessionist claims as a mechanism for dissipating tensions. Forms of limited self-rule may be introduced in an attempt to head off calls for complete secession. The decision by the British government to allow votes in Scotland and Wales on devolution in 1997 can be seen in the same way. While such decisions can be couched in terms of promoting democracy and encouraging a more people-centred approach, they can equally be interpreted as a means of countering the claims by Welsh and Scottish nationalists for total independence. In this way the state is maintained through the somewhat paradoxical method of reducing some of its own powers (Bogdanor 1999; see Chapter 7).

The case of regional autonomy in Spain provides another example of this response. The Basque country and other regions within Spain were granted limited forms of self-rule in part at least to try and quell pressures for com-plete secession. Spain's attempt to devise a constitution to reflect its internal divisions demonstrates the tension concerning limited autonomy within states. Although forms of self-rule have been granted to regions within Spain, the 1978 Spanish constitution re-affirms the centrality of the Spanish state. Article two states that:

> The constitution is founded on the indissoluble unity of the Spanish nation, the common and indivisible *patria* of all Spaniards, and recognises and guarantees the right to autonomy of the nationalities and regions integrated in it and the solidar-ity among them. (cited in Guibernau 1995: 245)

In this way, there is the assertion of the apparent ambiguity of a Spanish nation with other nations embedded within it. A similar tension occurs in Russia whose 1993 constitution affirms the right of national self-determination but then expresses the inviolability of the Russian Federation, thereby preventing its component nations from seceding (G. Smith 1995b).

One important point should be made about the above arguments. In the main, these pressures from secessionist nationalists and from the periphery might be seen as threats to the existence and legitimacy of the **existing** state, or to the extent of the state's territorial reach, but they are not necessarily suggestive of the demise of the state as a **concept**. In most instances, it is a case of replacing the existing state by one or more others. In other words, it is not

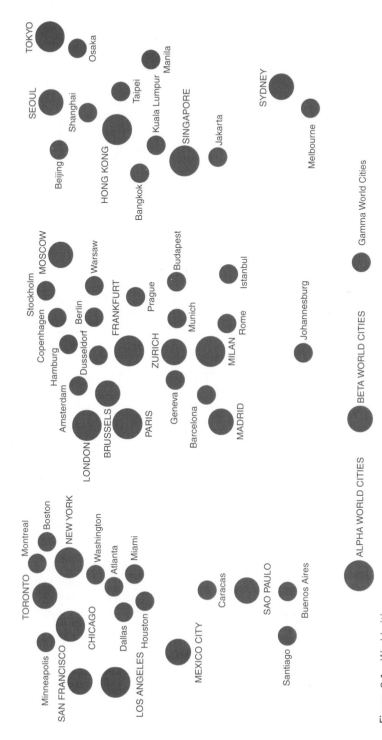

Figure 6.1 World cities.
Source: Taylor and Flint, 2000

the idea of the state that is in question, merely the right of the current one to govern particular territory. A more serious threat to the future of the state as a political–territorial formation is presented by external or trans-state pressures. It is to these that attention is now turned.

Pressures from above

In addition to internal pressures on states there are also what can be seen as external threats, or pressures from above, to the viability of the state. These tend to result from the processes commonly placed under the umbrella of globalization. This refers to the view that we live in an increasingly interdependent world where knowledge of other people and places is constantly available to us all and where there are regular interactions between people in one place and those in other places. Anthony Giddens has defined globalization as 'the intensification of worldwide social relations which link distant localities in such a way that local happenings are shaped by events occurring miles away and vice versa' (1990: 64). There is, thus, a stretching and deepening of social relations across space. In effect, the world is being made a smaller place through a series of mechanisms and processes. Economic transactions, telecommunications advances, information technology, television, cheaper and quicker air travel are all bringing previously 'remote' places closer to us. Indeed, the world could be said to be brought into our very living rooms on a daily basis. Some geographers have likened these globalizing tendencies to time–space compression; in an increasingly interdependent world, fashion and other trends diffuse quickly around the globe and permeate political and cultural boundaries quickly and easily (Harvey 1989; Massey 1994). The rise of the internet and the 'informational society' is seen to render the world an increasingly smaller and smaller place. In short, it is suggested that we live in a global world in which there are expanding opportunities for interaction and in which, as a consequence, territory and borders are of much diminished significance. It has been argued that key 'world cities' are assuming greater importance than individual states in the global era (Figure 6.1). All of this suggests that the concepts of nation and state are becoming increasingly irrelevant as we move towards an ever more cosmopolitan world in which questions of nationality and place become less and less important.

Various distinctive elements within the globalization process can be identified. These are (derived from Held 1989):

1. growth of international trade, capital flows, investment, etc.;
2. improved communications and technology;
3. transnational organizations (TNCs, etc.);
4. international institutions (UN, IMF, World Bank);
5. regional economic blocs (EU, etc.);
6. defence and military alliances (NATO);
7. processes aimed at harmonizing international law;
8. cultural diffusion (spread of English language, tourism, etc.).

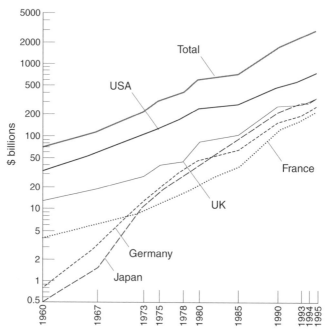

Figure 6.2 Growth of foreign direct investment, 1960–95.
Source: Dicken, 1998

These various elements can be subsumed under the broad headings of economic, political, cultural and environmental dimensions. However, it is important to bear in mind that these are not discrete categories. Most social scientists accept the interconnectedness of these elements; the economy cannot be viewed in strict isolation from society; culture cannot be detached from environment and so on. Indeed, one of the central elements within globalization is the emphasis on the interrelationships of these elements. Nevertheless, for purposes of explanation, these are briefly discussed below.

Economic dimensions

It is within the economic arena that many of the most obvious signs of globalization are manifested. There have been sizeable increases in the extent of international trade and in the growth of foreign direct investment (Figures 6.2 and 6.3). The importance of overseas holdings for banks has grown dramatically. Two key manifestations of globalization are increased interstate economic co-operation and transnational corporations (TNCs), both of which call into question the sovereignty of the territorial state. Growing economic co-operation is reflected in the growth of suprastate institutions. These are regional amalgams of (usually) neighbouring countries engaged in formal economic or political alliances (Figure 6.4). The European Union represents the best-known example of this (Figure 6.5). Formed initially in the late 1950s as an economic union of six countries (Belgium, France, Italy, Luxembourg,

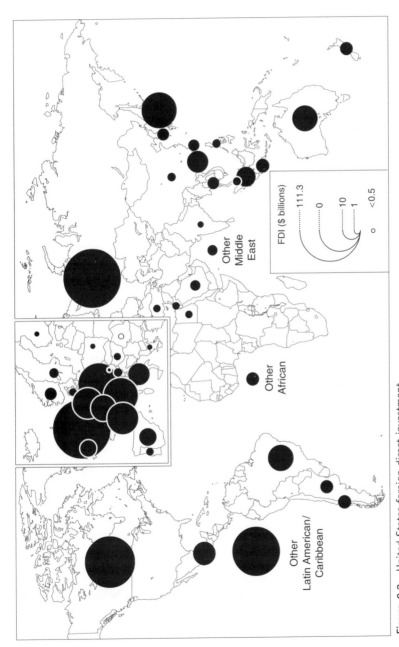

Figure 6.3 United States foreign direct investment.
Source: Dicken, 1998

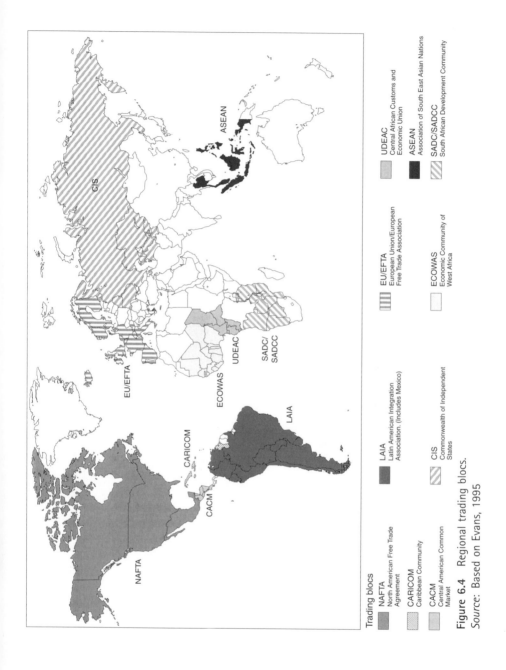

Figure 6.4 Regional trading blocs.
Source: Based on Evans, 1995

Trading blocs

NAFTA
North American Free Trade Agreement

CARICOM
Caribbean Community

CACM
Central American Common Market

LAIA
Latin American Integration Association: (Includes Mexico)

CIS
Commonwealth of Independent States

EU/EFTA
European Union/European Free Trade Association

ECOWAS
Economic Community of West Africa

UDEAC
Central African Customs and Economic Union

ASEAN
Association of South East Asian Nations

SADC/SADCC
South African Development Community

Figure 6.5 European Union.

Netherlands, West Germany), it has now evolved into a more overtly political arrangement and membership has risen to fifteen countries with a waiting list of applicants. In some respects the EU is beginning to take on the appearance of a 'superstate' in which power is taken away from the constituent members and is vested in Brussels. Similar trading blocs exist within other parts of the world. These include the Association of South East Asian Nations (ASEAN) and the North American Free Trade Agreement (NAFTA); however, none is as well developed as the EU (see Box 6.2).

A second economic threat to the sovereignty of the nation-state is the increasing economic power wielded by large businesses. TNCs are large companies which operate in two or more countries and they can exert considerable influence over government policy. They operate through a series of

Box 6.2   European Union

The Treaty of Rome in 1957 established the European Economic Community (EEC) which has evolved into what is today known as the European Union (EU). Its original aims were broadly economic, centred on the creation of a common market between the six founding members: France, West Germany, Italy, Belgium, Luxembourg and the Netherlands. Since its formation there have been a number of additions which have seen the union expand territorially. The UK, Republic of Ireland and Denmark acceded in 1973 followed by Greece in 1991, Spain and Portugal in 1986 and Austria, Sweden and Finland in 1995. Currently, there is a waiting list of eastern European countries who, following the demise of communism, see their economic future increasingly tied to the west.

Just as there has been territorial expansion, so too has there been a political and economic deepening of the union. From its original focus on economic co-operation there is now an increasing emphasis on political co-operation (Table 6.1). Attempts to move towards a harmonization of aspects of foreign policy and immigration policy are two examples of this. The spatial expansion and political deepening of the union have been resisted in some quarters. In

Table 6.1   Deepening of the European Union

Year	Key developments
1951	Treaty of Paris: formation of ECSC: Belgium, Netherlands, Luxembourg, France, West Germany, Italy
1957	Treaty of Rome: formation of EEC and Euratom
1960	Formation of EFTA: UK, Spain, Portugal, Switzerland, Norway, Sweden, Finland, Denmark, Ireland, Iceland, Liechtenstein
1967	Merger of EEC, ECSC and Euratom
1973	Accession of United Kingdom, Ireland and Denmark
1979	European Monetary System is established; introduction of ECU
	First direct elections to European Parliament
1981	Accession of Greece
1985	Publication of White Paper setting out details of single market
1986	Accession of Spain and Portugal; Signing of Single European Act
1989	Delors Report: blueprint for monetary union
1990	Incorporation of the former German Democratic Republic
1991	Treaty on European Union (Maastricht Treaty)
1993	Completion of single market
	Foundation of European Union to succeed European Community
1994	Extension of single market area to include EFTA
1995	Accession of Sweden, Finland and Austria
1999	Launch of single currency

(continued)

(continued)

most countries there is opposition to what is viewed by some as a diminution of national sovereignty as power is seen to be taken out of the hands of national governments and handed to Brussels. The EU is thus the most evolved example of interstate co-operation, one which might be seen as providing a threat to the continued pre-eminence of the state as a political unit.

Currently the EU has introduced a rudimentary form of European citizenship. Following the signing of the Maastricht Treaty (Treaty on European Union) in 1993, citizens of EU member states automatically became EU citizens. European integration proceeds both through supranational institutions such as the European parliament and also through intergovernmental co-operation over specific issues. Bodies such as the European Court of Justice, based in Luxembourg, now have a certain degree of power over individual states. Most recently, January 1999 saw the first stage of the introduction of the Euro – a single currency for the European Union. As yet, the currency is not in circulation but the eleven signatory members are now in a currency union whereby each member-state's currency is tied to a set exchange value to the Euro and, hence, to each other. The Euro itself will come onstream in 2002. Such a move represents a weakening of economic power for individual states. In the UK, at the time of writing, an anti-Euro campaign is getting under way aimed at keeping the UK out of this currency union.

Despite this apparent usurping of state power by an external body, some such as Anthony D. Smith (1995) argue that national loyalties and national cultures will remain predominant, principally because there is no pan-European culture to replace them. Even with the introduction of the Euro, each participant country will use its own 'national' symbol on its notes and coins. While the EU may take power from individual states, national loyalties seem unlikely to diminish significantly in the near future.

branch plants, the locations of which are determined by the relative differentials in wage rates, transportation costs, government grants, infrastructure, proximity to markets, etc. It is estimated that some 39,000 parent companies control 265,000 foreign affiliates (Dicken 1998). A sizeable amount of world trade is accounted for by these TNCs and a considerable proportion of this is trade between branch plants of the same company. This accounts for up to one-quarter of world trade (Goldblatt *et al.* 1997). These global corporations provide employment for many millions of people and their presence in particular countries is often seen as highly beneficial in terms of raising economic output, employment levels and wage rates. In many countries the amount of employment accounted for by branch plants of TNCs is extremely high. Many of the world's largest corporations have a considerably higher turnover than many 'Third World' countries. This, and other factors, have led to considerable criticisms of TNCs for operating effectively as neoimperialist institutions. Whatever the view taken on TNCs they undoubtedly display something of a global reach which is relatively unhindered by territorial boundaries.

Political

In the political sphere, the end of the Cold War has reputedly led to the victory of social democracy and has ended, or at least diminished, the ideological divisions of the past. This is the argument made popular through the publication of Francis Fukuyama's *The End of History* (1992). The collapse of communism and the subsequent encroachment of capitalism and liberal democracy into eastern Europe are seen as homogenizing forces leading to the end of ideology and, hence, it is asserted, the end of major conflict.

At another level the increasing evidence of political co-operation is taken as symptomatic of the decline of the importance of single-state action. Thus, in the moves towards political unity in Europe, or through the activities of the UN, there is seen to be a trajectory towards increasing political co-operation; for example in the struggle against drug trafficking or against 'terrorism', as conventionally defined. NATO's war against Yugoslavia in 1999 might be viewed as a form of suprastate policing. Related to this, the evolution of systems aimed at harmonizing international law, the creation of various international legal protocols, the International Court of Justice in The Hague and a series of related developments can be seen as evidence of the creation of sets of supranational instruments which operate across territorial boundaries and which internationalize politics.

In addition to these formalized versions of political integration and co-operation, there are a number of what might be described as alternative political visions which call into question the continuance of a world state system. To some extent these incorporate elements arising from below, i.e. within the borders of existing states, together with philosophies and visions which transcend state boundaries. Anarchist thought suggests a world free of formalized territorial divisions where borders do not exist. In a similar vein, variations of socialist thought suggest alternative visions on the nature and functions of states. Perhaps more significant at the moment are social movements built around alternative issues such as the environment or around various politics of identities, as discussed below. Such visions call into question the whole notion of borders and formalized territorial divisions. Alternative visions promote notions of co-operation rather than interstate rivalry and they tend to envision a future whereby 'irrelevancies' of place of birth and national affiliation are seen as meaningless in a world dominated by a common humanity.

Cultural

In the cultural sphere people are wearing similar clothes, listening to the same types of music, reading the same literature (albeit, perhaps, in translation) and sharing similar values. The ubiquitous presence of CNN, MTV and other global media in hotel rooms around the world both reflects and sustains this process. While cultural differences persist, it is undoubtedly the case that a certain uniformity is becoming apparent. Step off a plane in Birmingham, Brussels or Bangkok and a McDonalds or BurgerKing is never too far away –

the 'Golden Arches' of modern civilization beckon. The growing importance of a small number of languages, most notably English, as the international languages of business and politics serves to further underline this process of cultural homogenization. Tourism, the world's fastest growing industry, is another element which can be seen as a contributor to the flattening out of cultural differences (Urry 1990).

Linked to all of this is an increasing emphasis on other forms of identity. Groups mobilized around issues such as gay rights, women's rights, etc. elevate such questions of personal identity above concerns with national identity. In this way, cross-cultural solidarity and the growing significance of a politics of identity transcend political borders. This has resulted in considerably greater cultural mixing and the development of what might be seen as cross-national and cross-cultural solidarity in ways which undermine traditional borders.

Environmental

Since the 1960s there has been an upsurge of interest in the environment and, in particular, a growing concern with the extent to which it is being destroyed. The environmental consequences of industrialization, increased car use, burning of fossil fuels and other activities seen as wasteful or harmful have contributed to the growth of an environmental movement and the development of green politics. The growing problems of pollution, whether of water, air or land, are seen as ones which can only be resolved by international co-operation. Viewed in this way, a world of sovereign nation-states may act as an impediment to resolving environmental problems. Some green visions of the world see territoriality in terms of a unified world which does not belong to anyone but which is held in trust for future generations. Because pollution does not recognize borders, a 'green' vision transcends state boundaries and employs a very different notion of territoriality.

Do states still matter?

Globalization has obvious implications for the world of nation-states. If globalization is occurring, then what is the future for independent political entities? In a nutshell, we live in an increasingly interdependent world where, it might be supposed, national boundaries are of ever-diminishing importance owing to the processes outlined above. However, a number of qualifications need to be made here. Despite the undoubted sense in which the world might be seen as experiencing globalization, much debate revolves around the issue.

In essence four key points can be made. The first is that globalization, in the sense of linking places together, has been occurring for a considerable period of time. Secondly, globalization is not experienced by everyone, or at least not in the same way. Thirdly, globalization is actively resisted in many diverse ways and, finally, even within a globalizing world, it can be argued that states and borders will continue to serve particular functions and will, as a consequence, endure.

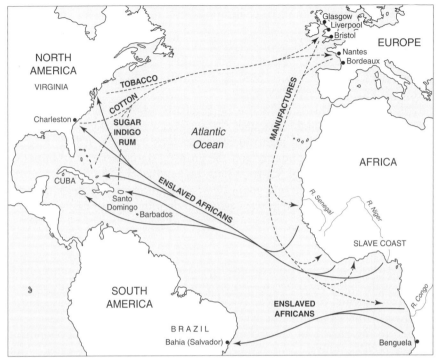

Figure 6.6   Triangular trade.
Source: Based on Duignan and Gann, 1985

Globalization is not new

The phenomenon of an integrated world in which the fate of particular places is linked to events in other, perhaps quite distant, places is not simply a feature of the later twentieth century. Trade and commerce have always served to link people and places and the 'global economy' is one which has evolved over time (Knox and Agnew 1994). Imperialism was a project which linked places together – not necessarily to their mutual advantage. The 'triangular trade' involving European countries and their colonies, whereby raw materials and manufactured goods made their way to and from Africa and the Americas and people were shipped as slaves from Africa to North America and the Caribbean, represented a form of globalization whereby many diverse places were integrated into an increasingly global economy with social and cultural implications (Figure 6.6). The East India Company was one of the TNCs of its day. Within the broad realm of culture, there has been continual inter-change of ideas, styles and modes of behaviour. Music, as one element of culture, has continually evolved through processes of cultural interchange. In brief, it might be argued that globalization is occurring at a much faster rate than heretofore, but the phenomenon itself is not new.

Plate 6.3   Dwelling beneath a bridge, Bangkok, Thailand.

Globalization is not universal

If globalization is not new, then neither is it universal; not everyone experiences it or at least not in the same way. Despite the apparent globalization of economy, society and culture, many people:

> may know little of the world beyond their own village or neighbourhood, even as their lives and localities are stunted or turned upside down by forces operating beyond their local horizons. Not for them the 'brave new world' of high-tech globe-trotting and jet-setting, more likely the fear or reality of unemployment, starvation or war. (Anderson, Brook and Cochrane 1995: 1)

The world may be becoming a smaller place for academics and business people on transatlantic flights, as Massey (1994) has argued, but it may be remaining much the same for those on the mid-Atlantic islands over which they fly or for those sleeping under bridges in air-route hubs such as Bangkok (Plate 6.3). Similarly, not everyone lives in a world of CNN and MTV – only those with the time and money or social position which allow or encourage them to do so. Despite the undoubted rapid advances in the speed at which money, fashion, information and ideas travel around the world, and despite the evidence of vastly increased cross-border interaction, this picture of an ever more global world is not the complete story. While globalizing processes are occurring, there are many who remain locked into their own local worlds. Viewed in this way, it is obvious that geography, territory and place continue to matter. It should of course be pointed out that being locked into a local world does not mean that the outer world has no impact. First world companies can, through their decisions on where to locate, their wage levels and their employment conditions, determine the quality of life for people in places such as Vietnam and Thailand.

Resistance to globalization

Despite the obvious globalizing tendencies outlined above, it is equally clear that these processes are constantly being resisted. Attempts at homogenization appear to always be met by attempts to resist and to assert difference. There are numerous examples of both overt and less obvious resistances to the flattening out of difference which is allegedly taking place. Within each of the arenas of globalization outlined above we can see examples of resistance to the globalizing process.

While TNCs undoubtedly wield enormous power, considerable opposition to their activities is evident particularly in many less developed economies where their power to operate without adequate environmental safeguards, to utilize non-union workforces and other practices are quite often resisted by local pressure groups. Their ability to dictate their own conditions and to open and close their operations in different countries is seen to lead to problems in terms of factors such as sustainable employment. The transitory nature of branch plants, derived from TNCs' ability to relocate relatively easily, is pointed to as evidence of the dangers attaching to a dependence on such enterprises as a mechanism for economic growth or employment generation. In the view of many, the power wielded by TNCs is antithetical to the interests of many states and to sections of the population in particular countries. Disasters such as that which occurred at the Union Carbide plant in Bhopal in India serve to highlight the dangers and provide a focus for opposition. Protests against Shell's activities in Nigeria and its alleged collusion with the Nigerian government in suppressing opposition provide a further example of resistance to a global corporation, utilizing in this instance the power of local identity (see Box 6.3).

Within the political sphere, the present on-going process of European integration has taken on significant political dimensions and this has led to objections from certain quarters that national identity and sovereignty are being lost. In Britain opposition to the continued economic and political integration of the EU coalesces around two broad perspectives. Those on the right of the political spectrum view current developments in terms of a loss of national sovereignty; a process which, it is suggested, is diminishing Britain as a nation and which is eroding its right to act in its own national interests. 'Outsiders' are increasingly seen to be interfering in matters deemed to be rightfully British. For many of those on the left of the political spectrum, closer integration carries the threat of a diminution of democracy. The EU is seen as a largely undemocratic body which is far too removed from the inhabitants of the member states. Current processes are seen as leading to the creation of a 'banker's Europe' rather than a 'people's Europe'. In any event, power is seen to be taken away from individual member states and handed over to a central authority. Such moves have been, and continue to be, resisted in various

---

Box 6.3   TNCs and the state: the case of Shell in Nigeria

Nigeria gained formal independence from Britain in 1960. The country is composed of a collection of numerous different ethnic groups many of whom feel marginalized within the Nigerian state. Since independence, the country has had a somewhat chequered political history of corruption, military regimes and unrest. The regime of General Abacha between 1993 and 1998 was one which dealt harshly with political dissidents. The Shell oil company has operated in Nigeria since the 1930s and has extensive operations there, particularly in Ogoniland. Its activities here have been highly controversial and have engendered considerable opposition from local Ogoni people and from a range of activists concerned about Shell's environmental record. Oil production accounts for around 90 per cent of Nigeria's foreign exchange receipts and constitutes over 70 per cent of government budgetary revenue (Khan, 1994). It is also estimated that 14 per cent of Shell's global production is from Nigeria. The relationship between state and company is clearly mutually beneficial. The picture for many of the state's citizens is, however, somewhat different.

It has been claimed that there is a close relationship between the military regime and Shell with allegations of payments made to enhance the security of Shell installations. While the company denies any wrong-doing, and, while specific allegations have been denied, it has been argued that 'Shell has been happy to maintain an unjust and corrupt political system which enables the company to disregard environmental standards and to keep wage costs and social overheads to a bare minimum' (Gilbert, 1999). Criticisms of Shell came to a head with the judicial execution of the writer and activist Ken SaroWiwa – who had been very critical of Shell's activities – and eight others on charges of murder in 1995. It is felt by many that Shell could have prevented this and that the company is happy to profit from the oppressive policies of the state. In any event Shell's power in Nigeria provides a useful illustration of the influence which large corporations are capable of wielding. As an interesting footnote to this issue, it has led to a split within the Royal Geographical Society, one of whose sponsors is Shell. A group called the Critical Geography Forum emerged in the aftermath of the decision of a number of members to resign from the RGS because of their unease with Shell's sponsorship.

---

quarters. In Denmark, the Single European Act of 1986 was only narrowly ratified, and even then only after a second referendum, having been defeated the first time around. Norway has also declined to join the EU on a number of occasions. Such resistances, of course, may be based on pragmatic evaluations of economic prospects rather than any 'pure' nationalist sentiment.

While many point to the end of the Cold War as ushering in a new era of global co-operation, it is readily apparent, as made clear in the previous chapters, that national identity and fragmentation may well be the order of the day. The continuance of ethnic and nationalistic tensions in many parts of the world, most notably in eastern Europe, suggests that a world without borders is a long way off. The substate nationalism discussed in the previous section contradicts in many ways the idea of a global borderless world. Borders seem likely to continue to exist. Of course, this does not mean that the power of states remains the same. Diminutions in the extent of state sovereignty and effective alterations to the powers of states may well occur.

The persistence of sizeable global inequalities is sure to reinforce opposition to the forces of homogenization. Re-assertions of interest in local languages, customs and cultural practices, although they may become closely bound up with tourism and the heritage industry (themselves symptomatic of a globalizing tendency), do nevertheless represent attempts at asserting or re-asserting a distinctive cultural identity and place distinctiveness amidst the apparent homogenization of the turn of the century.

Are states still necessary?

Another significant point must also be made. Although national borders have become increasingly more permeable, that does not mean that they have ceased to have relevance or that they are not needed. Even in the recognition of the immense power wielded by TNCs, it should be borne in mind that many of their activities are dependent on the existence of states. These include the propensity of TNCs to move from country to country, thus producing instability in employment levels and so on. Economies may become very heavily dependent on these corporations. In order to keep TNCs, countries may be prepared to alter or weaken employment or environmental legislation, thus being effectively 'blackmailed' by these powerful non-governmental institutions. In this way, nation-states may be said to be ceding control to supranational private sector institutions. However, it is equally obvious that TNCs require the state system. They could not engage in playing off one country against another if those countries did not exist in the first instance.

Even within the EU, while borders are disappearing, an external perimeter fence is being erected making entry to the member states increasingly more difficult. There is an increasing emphasis on free movement of goods, capital and labour across the borders of the member states. However, this move towards equal internal access is accompanied by an increasing 'hardening' of the external borders of the EU. Entry of particular goods, such as certain foodstuffs, may be met with prohibitive common tariffs while moves towards harmonizing a common immigration policy have been gradually evolving. This has given rise to the notion of 'Fortress Europe' whereby internal free-dom of movement is matched by an external perimeter fence which will become increasingly difficult to cross, particularly for those lacking skills and, by implication, those from poorer countries. As a consequence one set of borders may be disappearing while another is being simultaneously reinforced. In the words of former UK prime minister John Major 'we must not be wide open to all-comers just because Rome, Paris, and London are more attractive than Bombay or Algiers' (quoted in *Observer*, 30 June 1991). It might be argued that the evolution of regional superstates merely reflects a change in the present configuration of states, not a fundamental change in the idea of territorially based state control.

While the alternative political visions outlined earlier illustrate resistances to a state-centred world and suggest moves away from this framework, some cautionary words are necessary. The concerns enunciated above (environmental issues, gay rights, etc.) are more often than not articulated within a 'nationalist' framework. In most countries there are national groupings or national branches of international groups concerned with these issues – national branches of Amnesty International, Greenpeace and the like. They explicitly recognize the world of the nation-state and, while they do not necessarily conform to a state-centred way of thinking, they nevertheless see current territorial structures as a convenient organizational device. Even in attempts to proclaim the irrelevancy of territorial divisions, the methods used can be explicitly territorial, as for example in the old Campaign for Nuclear Disarmament tactic of declaring specified areas nuclear free.

Rather than taking over from national identity, it may well be that these questions of personal identity will continue to coexist in a system of overlapping identities rather than simply replacing national identification. Thus, one can be female and/or homosexual and/or black but also Italian. Which identity takes priority at any one time will depend on the specific context in which the individual finds herself. In any event, whether or not national identities retain pre-eminence, it is abundantly obvious that there are strong resistances, many of them quite localized and territorially based, to processes of globalization and homogenization. Just as there are geographies of global integration so too there are geographies of resistance.

Finally, the above debates may tend to suggest that we are living through something of a transitional phase whereby current territorial structures are about to be replaced by new forms of territorialization or by non-territorial methods of organization. However, James Anderson (1995, 1996) argues that, rather than seeing this as a transitional stage as we move away from the nation-state, it may instead represent a new stage in itself. Anderson suggests that, perhaps, this is a 'stable' form rather than a period of post-state instability. Pressures of globalization on the one hand, and the need to respond to ethnic and regional groupings on the other, may be creating a postmodern set of overlapping sovereignties more reminiscent of the medieval era than the modern one. As Anderson further points out, external pressures have always been felt by the politically less powerful states, most notably colonies and former colonies. Simply because major powers or ex-powers, such as Britain, are now feeling versions of this does not in itself herald the end of the nation-state. In conclusion, therefore, it can be argued that the world-state system is far from dead. Despite, perhaps because of, globalization, states and state boundaries seem likely to continue to function as significant elements in our everyday lives well into the future. In addition there is no huge sign that national allegiances, as distinct from state power, are being significantly diminished.

Summary

This chapter has outlined two types of pressure said to be threatening the world of states as currently constituted. Firstly, there are pressures from below seen to be associated with internal divisions within states centred along fault lines such as nationality, spatial inequalities and language differences. Secondly, there are pressures from above largely associated with forces of globalization operating in the economic, political, socio-cultural and environmental spheres. While these various pressures undoubtedly exist, they will not necessarily lead to the demise of the state. Most internal pressures centre around the replacement of the current state with one or more new ones. Equally, many of the globalizing tendencies outlined above are actively resisted, suggesting that national perspectives and frameworks will continue to be of major importance. All of this suggests that rumours of the state's imminent demise have been greatly exaggerated (Anderson 1995) and that place, and with it contestations over territorial control, will definitely continue to matter.

Further reading

On internal pressures on states see Richard Muir, *Political Geography. A New Introduction* (Macmillan, Basingstoke, 1997) and various articles by J.B. Forrest including 'The quest for state hardness in Africa', *Comparative Politics* 1988, **20**: 423–41. For a discussion of globalization, see the volume edited by Mike Featherstone, *Global Culture: Nationalism, Globalisation and Modernity* (Sage, London, 1990), Anthony Giddens, *The Consequences of Modernity* (Polity Press, Cambridge, 1990) and the three-volume series by Manuel Castells, *The Rise of the Network Society, The Power of Identity* and *End of Millennium* (Blackwell, Malden MA, 1997). On the various pressures on the state, see Anthony D. Smith, *Nations and Nationalism in a Global Era* (Polity Press, Cambridge, 1995) and *A Global World? Re-ordering Political Space*, edited by J. Anderson, C. Brook and A. Cochrane (Open University/Oxford University Press, Oxford, 1995), particularly the chapter by Anderson, 'The exaggerated death of the nation-state'; see also Anderson's article 'The shifting stage of politics: new medieval and postmodern territorialities?', *Environment and Planning D: Society and Space* 1996, **14**(2): 133–53. For broader discussions of globalisation and time-space compression see David Harvey, *The Condition of Postmodernity* (Blackwell, Oxford, 1989). See also Francis Fukuyama *The End of History and The Last Man* (Free Press, New York, 1992). Peter Dicken's *Global Shift. Transforming the World Economy* (2nd edition, Paul Chapman, London, 1998) provides an exellent discussion of current global economic–geographic trends.

On the European Union see relevant chapters in T. Unwin (ed.), *A European Geography* (Addison Wesley Longman, Harlow, 1998) and in Brian Graham (ed.), *Modern Europe. Place, Culture and Identity* (Arnold, London, 1998).

On the Basques see R. Collins, *The Basques* (2nd edition, Blackwell, Oxford, 1990). For a discussion of Belgium's language divide see the various articles by Alexander Murphy including 'Linguistic regionalism and the social construction of space in Belgium', *International Journal of the Sociology of Language* 1993, **104**(1): 49–64, and 'Belgium's regional divergence: along the road to federation', in G. Smith (ed.), *Federalism. The Multiethnic Challenge* (Longman, London, 1995).

On environmentalist thought, green politics and the growth of the environmental movement see David Pepper, *Modern Environmentalism: An Introduction* (Routledge, London, 1996), Andrew Dobson, *Green Political Thought* (2nd edition, Unwin Hyman, London, 1995) and T. Doyle and D. McEachern, *Environment and Politics* (Routledge, London, 1998). On the issue of Shell in Nigeria see the articles in *Ethics, Place and Environment* 1999, 2(2).

Chapter 7

Substate territorial divisions

So far in this book the main concern has been with territories and territorial behaviour at a macro-scale centred on discussions of nationalism, state formation and interstate territorial relations. However, as indicated earlier, territorial behaviour and territorial strategies can be seen to operate at much smaller spatial scales. Some aspects of this, those surrounding substate nationalisms, have been addressed in the previous chapter. This chapter and the one which follows focus on other expressions of territoriality and the utilization of territorial strategies at a substate level. In this chapter the focus is on formalized territorial divisions beneath the level of the state. The following chapter deals with what might be seen as informal territories of a somewhat more ephemeral nature, focusing on issues such as racialized space and gendered space. However, as is shown in the present chapter, the utilization of features such as ethnicity in delineating places is sometimes legitimized through formalized territorial strategies.

The initial concern in this chapter is with internal territorial divisions which governments or their agencies use in order to 'manage' the state. Notwithstanding predictions of the demise of the state, as outlined in Chapter 6, states continue to employ internal territorial strategies in order to govern more easily. This is generally done through the creation of substate geographic units within which local or regional bodies have some form of control. These internal territorial structures may range from being relatively powerless and weakly developed, as in the case of highly centralized states, to being very powerful, as in the case of federated or highly devolved states. This chapter initially considers these formal state territorial divisions. In addition to these divisions, there are a range of quasi-state or semi-state bodies with particular responsibilities in relation to their designated areas. These can be seen as arms of the state with regional responsibilities. These include regionally based health authorities, police forces and the like. Together with local and regional governments, these various agencies can be said to constitute the local state: a set of social and institutional relationships existing at a local level.

Within countries such as the United Kingdom, the programme of 'rolling back the state' has led to an increasing reliance on the private sector to provide services previously seen as the preserve of the state or its regional organs. The increasing numbers of local service providers mean that the local state is becoming ever more fragmented, leading to a consideration of modes of governance rather than government. Associated with these recent changes in how services are provided is an increasing emphasis on the role of local territorially based communities. Community-based organizations who represent the interests of residents within defined localities are increasingly being drawn into the formal political arena and are important agents in creating and sustaining a territorial identity. All of this means that these various agents (commercial and voluntary) are becoming part of new territorial arrangements in which the division between the public and private sectors is becoming increasingly blurred. Central to this chapter is a concern not so much with documenting the specifics of territorial arrangements, but rather emphasizing the manner in which space is used in order to administer and control. As with the examples in previous chapters, these internal territorial arrangements represent a spatial expression of power; a means through which the state or its agencies manage their affairs.

Formal political–territorial subdivisions

Within most countries there are internal substate political–territorial divisions. These are geographical subunits which enjoy some limited powers over their own affairs. The extent to which states devolve power to local territorial forma- tions is quite spatially variable. Some states have highly devolved political sys- tems while others are highly centralized. Moreover, within individual countries the extent of autonomy accorded to substate territorial units alters over time.

At a basic level a differentiation can be made between centralized states and federal states. In the former, power resides almost exclusively with central government with only a minor role, if any, for regional or local bodies; in the latter, various powers may be devolved to regional or provincial governments. In a federal state, such as the United States or Germany, regional governments have a considerable degree of autonomy and can pass local laws. In contrast, highly centralized states have a system in which the degree of local devolution may be quite minimal. Obviously, many countries adopt political systems which exist on a spectrum between highly centralized and highly devolved. In the following sections, a brief discussion of federal states and local government is provided. While these spatial arrangements exist primarily for administrative purposes, it needs to be remembered that territorial divisions at this scale, just as with those referred to in previous chapters, may also inculcate a particular sense of place identity. The recent trend of rolling back the state in some western democracies is briefly discussed, followed by a theoretical appraisal of decentralization.

Federated states

The most devolved political–territorial systems are in federal states in which a significant amount of decision-making is carried out at a local level. Federalism is a political arrangement often favoured in multi-ethnic states. The ostensible aim is to ensure a reasonable balance between national and regional interests and (usually) between the various ethnic groups living within the confines of

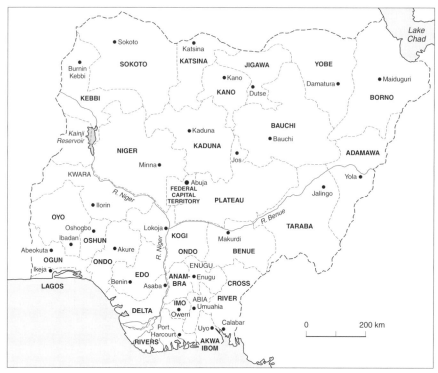

Figure 7.1   Federal units in Nigeria.
Source: Dent, 1995

the state. Federal states can be seen, in many respects, as offering a practical response to tensions which might otherwise undermine the state's continued existence. Ethno-regional groupings are accorded power over their region within an overall federal system. In some instances complex systems have evolved to try and deal with a multiplicity of ethnic, religious and/or regional tensions, as in India and Nigeria. In the latter case, more and more federal units have been created since independence (Figure 7.1). Some countries favour relatively large federal units, as in India, others favour very small units as in the Swiss *cantons*.

An obvious example of this is the Russian Federation which consists of a range of what can be seen as virtually separate countries bound together under the overarching authority of the Moscow government. Within the Russian Federation there are eighty-eight different types of federal unit, some of which enjoy greater autonomy than others (Figure 7.2). Ultimately what makes these territorial units something less than fully independent states, in the sense indicated in Chapter 3, is their lack of power in the arena of foreign policy and their consequent lack of complete sovereignty over their territory; they remain subject to Moscow and have no international recognition as independent sovereign entities. However, as Graham Smith (1995b) indicates, a small number of these units are *de facto* operating independently of Moscow.

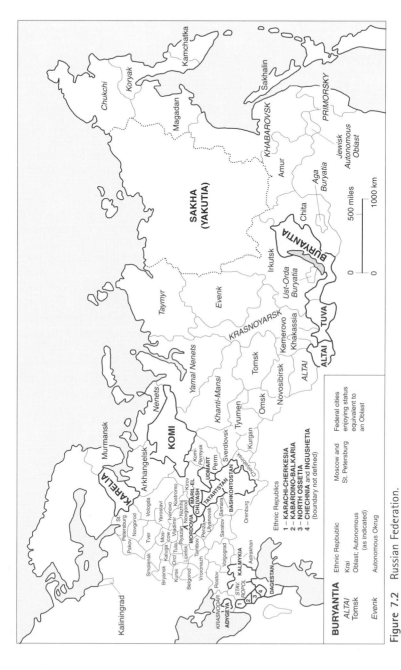

Figure 7.2 Russian Federation.
Source: G. Smith, 1995b

While it might be argued that federalism provides a solution to otherwise intractable ethnic conflicts, the experience of Chechnya, which, following a very bloody conflict in the early 1990s, effectively seceded from the Russian Federation, suggests that such an arrangement is far from stable, creating the potential for 'civil' war. The on-going tensions in the north Caucasus, involving Chechnya and Dagestan (referred to in Chapter 6), provide ample evidence that some of the autonomous regions in the federation are unhappy with current arrangements (Goldenberg 1994).

The United States provides a further example of a federal structure. Here, each state has its own governor and legislature and can implement its own laws on issues such as the death penalty and homosexuality. While each state has a range of powers, they all remain subject to the overarching political institutions of the federal government based in Washington and to which each state sends political representatives. Individual states do not have sovereignty; that resides with Washington. Spain is another example of a country employing a federal or quasi-federal mode of government; its seventeen provinces enjoy limited self-rule within the confines of the Spanish state in a somewhat asymmetrical system where the Basque country and Catalonia enjoy considerably more autonomy than the other provinces. As indicated in the previous chapter, this has not prevented calls for secession on the part of certain regions, most notably the Basque country.

The name of a country may suggest a particular form of territorial administration. However, there are states whose names suggest a highly devolved structure but where, in practice, power is tightly concentrated in the centre. An interesting example of this was the former Soviet Union. While it was composed of the Russian Federation and fourteen 'republics', it remained a highly centralized one-party state; a type of 'federal colonialism' (G. Smith 1995b). It follows that there is a need to distinguish between what may appear to be a highly internally territorialized system of devolved power and the reality in terms of effective devolution of power. In fact, the three communist federations of the Cold War era, the Soviet Union, Yugoslavia and Czechoslovakia, were all federations essentially imposed from the centre rather than being mutually agreeable solutions to internal difficulties. All three vanished rapidly, with the aftermath in Yugoslavia being particularly grotesque (see Chapter 5).

One very specific form of decentralization, which is of special interest in maintaining a particular form of political control, was that employed in apartheid South Africa. Apartheid (meaning apartness) was a policy formulated on the philosophy of racial segregation. The Group Areas Act (1950) and associated legislation effectively meant that people had to live in the areas designated for their 'racial group'. As a consequence of this, black people were excluded from living (although not from working) in many urban areas. This was a system designed to keep non-white people 'in their place'. White power was expressed territorially. One of its most obvious spatial manifestations was the creation of self-governing 'homelands'. These were referred to by South Africa as independent sovereign entities. In practice they were little more than internal regions with limited forms of self-rule (see Box 7.1).

Box 7.1   South Africa

South Africa's apartheid system was predicated on the belief that people could be categorized in terms of racial group; that the population could be divided into four broad categories: white, black, mixed race/coloured and Asian. This had a distinct spatial dimension. The white government decreed that blacks be allocated so-called 'homelands' or *bantustans*. Four notionally independent republics were created in the late 1970s and early 1980s – Bophuthatswana, Ciskei, Transkei and Venda (Figure 7.3). These were republics in name only, essentially remaining part of the Republic of South Africa and gaining no international recognition in the world beyond. In addition to the 'republics', six 'self-governing territories' were also created – Gazankulu, KwaZulu, KaNgwane, Lebowa, KwaNdebele, QwaQwa. These were supposed to evolve into republics but the dismantling of the apartheid state meant that this never occurred.

The Group Areas Act of 1950 legitimated racial segregation within urban areas, thus creating racialized territories at the urban scale (Figure 7.4). Black

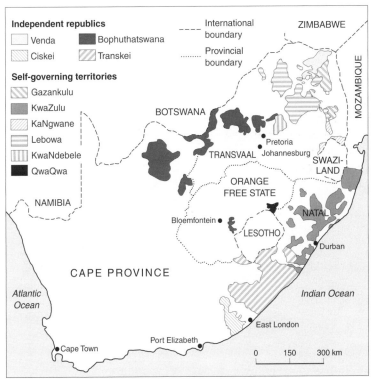

Figure 7.3   Territorial divisions in South Africa under apartheid.
Source: Smith, 1990b

(continued)

(continued)

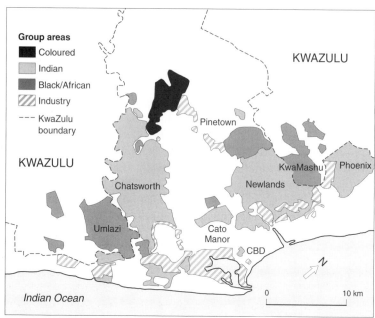

Figure 7.4 Group Areas and townships, Durban, South Africa, late 1980s. *Source*: Smith, 1990b

Plate 7.1 South African township. Photo courtesy of John Dixon.

(continued)

(continued)

workers were meant to reside in their 'proper' homeland but were needed to work in the 'white' cities. Because of this, huge townships developed providing 'temporary' housing for blacks (Plate 7.1). The 'races' were kept apart and a huge segregation of space occurred in pursuit of this racist ideology. While the rhetoric suggested that this policy of apartness was for the benefit of all South Africans, in reality it was a means of white control of the vast majority of the surface area of the country. Whites only constituted 14 per cent of the population but had control of 87 per cent of the land area.

Local government

In contrast to truly federated systems there are many examples of highly centralized states where there is little devolution of power and where central government exercises tight control over all parts of the national territory. However, even in these cases internal territorial subdivisions often exist. States still need mechanisms through which they can manage their sovereign territory. In the United Kingdom there is a complex system of local government. The situation is further complicated by the existence of different systems in England and Wales, Scotland and Northern Ireland. Even within England and Wales, there are different types of system.

Local government in the United Kingdom operates at different levels within a spatial hierarchy. Two major changes in the spatial framework of local government have taken place in the comparatively recent past. In 1973 many 'old' counties were replaced with newer metropolitan counties while others, such as Herefordshire and Worcestershire, were amalgamated. In 1998, there was a partial reversion to the older counties. Beneath the county councils are sets of district councils and throughout much of Britain a two-tier system of local government has operated with responsibilities split between county and district councils. Basically, items such as education, transport and social services tended to fall within the ambit of county councils while concerns such as housing, refuse collection, elements of local planning and recreational provision tended to fall within the remit of district councils. This situation has been made more complex by the 1998 changes which have resulted in the creation of a number of unitary authorities in some counties (such as the newly reconstituted Herefordshire) which combine the functions of county and district councils. In others, such as Worcestershire, the district tier remains (Figure 7.5).

Beneath the county and/or district councils lies another tier. Parish councils (of which there are over 8,000 in England) have a right to be consulted by district or county councils over issues affecting their area but they have very few powers of provision although they can provide or improve things such as village halls, car parks, camping sites and footpaths at their discretion. If the situation with regard to local government is complex with some areas of apparently overlapping responsibility between the tiers, the situation is further complicated by the fact that numerous quasi-governmental non-elected bodies

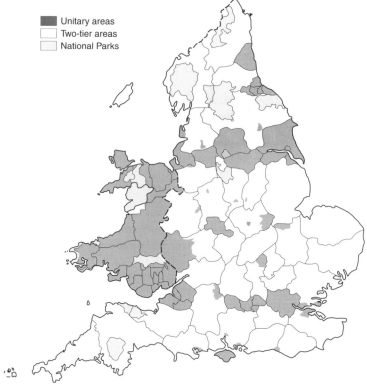

Figure 7.5   Planning authorities in England and Wales.
Source: Cullingworth and Nadin, 1997

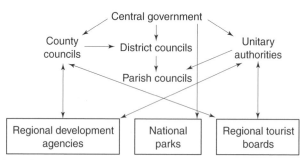

Figure 7.6   Simplified model of local government in England.

have responsibility for certain functions. These include local transportation authorities, health authorities and urban development corporations (Figure 7.6).

While the system of local government in the UK is largely based on historic divisions (partly ecclesiastical) of long standing, the French system of *départements* was constructed for administrative reasons after the revolution in the eighteenth century. At the time, this was an ahistoric attempt to create a 'rational' system

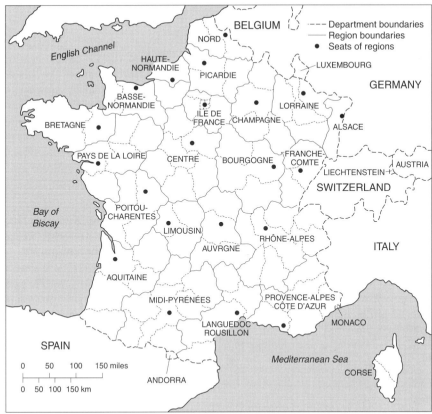

Figure 7.7 French *départements.*
Source: Glassner, 1993

of local government (Anderson 1996). The *départements* were instituted in order to form the territorial basis from which to administer the new republic. Although there have been changes in the degrees of power wielded at this level, the divisions themselves have endured to the present (Figure 7.7).

While many countries employ these internal territorial arrangements, recent trends in the provision of services at a local level suggest that the division between the public sector and the private sector is becoming blurred. The role of local authorities has altered substantially in recent decades. Central to this has been the diminishing role of the state as a service provider and the associated increase in the range of bodies who have some sort of role in local service provision. For many observers this represents a move away from more traditional forms of government.

Rolling back the state

The rise of 'new right' politics in the UK and elsewhere since the late 1970s, has resulted in a change in the way in which local government is conceived.

The key shift has been from provision to enablement whereby local authorities are now operating more as facilitators of service provision rather than delivering the services themselves. This shift reflects a free market ideology aimed at the minimization of state intervention. This has involved the de-nationalization of state companies and services and their subsequent privatization. Associated with this has been an increased tendency towards state agencies contracting out services. This policy was particularly favoured in the UK following the coming to power of the Conservatives under Margaret Thatcher in 1979. The policy of 'rolling back the state' might be said to be, in part at least, driven by the idea of saving government money through placing responsibility for service delivery in the hands of the private sector. However, while economic arguments were often used, there was undoubtedly a strong ideological component to this strategy. State intervention in the area of service provision was viewed in negative terms and there was a strong emphasis on individuality in preference to collectivism. The state was viewed as an entity whose authority needed to be reduced and whose influence on everyday life should be minimized.

It followed from the arguments outlined above that local government should be considerably reduced. Within the United Kingdom, the Conservative programme instituted by Thatcher, continued by Major and displaying no sign of reversal under Blair's 'New Labour' saw an erosion of local government power combined with the promotion of a different ethos. Whereas formerly local authorities were seen as providers, now they were to be facilitators of provision. Local authorities were encouraged to contract out services rather than necessarily providing them themselves. This switch is described as follows:

> Since 1988 the functions of central government have been reviewed systematically. The first consideration is whether the work needs to be done at all. If not, the function is abolished. Second, does the work need to be carried out by government? If not, the activity can be privatised. Third, where government must retain overall responsibility for the function, does it need to be provided by civil servants, or would the private sector have expertise to offer? (*The Civil Service Yearbook* 1994, cited in Cullingworth and Nadin 1997: 36)

In this way the role of both central and local government as service providers was considerably reduced. Instead, a range of quasi-state or non-state institutions became involved in the provision of services. Thus, housing associations, health trusts and the like have taken over powers of allocation and provision of specified services. The market place rather than the ballot box has become the medium of popular expression as the private sector takes over the role traditionally reserved for arms of the national or local state.

These moves have been heavily criticized by many. They are seen as representing a diminution of democracy by removing powers from elected local authorities and conferring powers on unelected quasi-autonomous non-governmental organizations or quangos as they are commonly known. Places are being governed via a range of extra-state institutions. While the democratic nature of such moves may be questioned, from a theoretical level they have sparked a consideration of how the state operates. In this mix of state, semi-state and extra-state provision some have argued for a consideration of ideas of governance. Hill (1994) suggests there has been a decentralization to individuals and communities while there has been an associated centralization of power into the hands of unelected quangos. These parallel processes in part could be seen as a de-territorialization of service provision on the one hand while, on the other, emphasizing the importance of territorially based communities and notions of empowerment. However, this seems to indicate primarily the power of individuals or households to decide who to buy their electricity or gas from, rather than power in the sense of political control. As Hill (1994) suggests there has been welfare empowerment rather than political empowerment.

Theories of decentralization

Local government can be placed within the same theoretical structure as our examination of the state in Chapter 3. Viewed from a liberal–pluralist perspective, forms of local administration can be seen as a set of mechanisms through which local needs can be ascertained and then met. In this way it might be argued that local government serves to bring about an acceptable solution to local problems. It might also be seen as facilitating the resolution of conflicting views at a local scale. The agencies of local government might be seen as providing a mediating filter between national interests and local interests on the one hand, and between competing interest groups at a local level on the other hand. In addition, local government can be viewed as a means of evening out the spatial blindness of overarching national policies through making them more sensitive to particular geographical areas and to local circumstance. Clearly, such internal territorial arrangements do render it easier to deliver services at a local level, while, on the political front, local or regional government (at whatever scale of devolution) can be seen to represent an attempt to deepen democracy by bringing the political process down to a local scale; it might be seen as furthering local control over particular issues. From this liberal–pluralist standpoint the local state is seen as a means through which the central state can be more responsive to local needs while providing a set of mechanisms through which greater public participation can be achieved. This also implies that such arrangements allow the state to operate in a more efficient manner.

To subscribe to such a view is to see local government and regional autonomy in purely functionalist terms; as merely being a mechanism for territorial management. However, it can be argued that such systems have a deeper significance than this. It is easy to regard devolved systems of power as a genuine attempt to democratize decision-making and to make regions more autonomous, but there may be a number of other factors leading to the creation of substate territorial units with some degree of autonomy. Following from theories of the state discussed earlier, it could be argued, from a structuralist perspective, that the territorial subdivisions are a mechanism through which the state can better organize itself. In order to retain its territorial integrity, the state needs to maintain its hegemony over all its geographical subareas.

Often this is best done through substate structures. Through internal territorial arrangements the state can ensure its legitimacy over all its geographic space. Even in highly devolved states, it can be argued that the ultimate aim is not greater democracy or sensitivity to local needs, but ensuring internal state hegemony. For example, devolution of power may serve to bring about the elimination or reduction of separatist pressures. Viewed in this way, the local state serves merely to bolster the central state and, in so doing, it enhances its ability to reproduce the existing political system and to further the interlinked processes of capital accumulation, labour reproduction and the maintenance of social order (through local policing, etc.). By being seen to involve local decision-making, the existence of local government also serves to further legitimize the state by promoting the veneer of greater democracy. This is not to argue that democratization is not occurring, rather it is to suggest that this is an 'accidental' outcome as opposed to the underlying purpose.

In Spain, as indicated earlier, there are seventeen autonomous regions dating from the 1970s, but based to a large extent on historical territorial formations. The regions were granted autonomy in the aftermath of the rule of the dictator General Franco who ran a highly centralized corporatist state in which attempts at local autonomy and expressions of cultural difference within Spain were suppressed. As previously suggested, autonomy for the Basque country has been a response to demands for full political independence by militant Basque separatists. Spanish political regionalism might be interpreted as an example of devolution practised in order to try and maintain and strengthen the state, not to weaken it. In this way, autonomy for the Basque country, Catalonia and elsewhere might be seen as an attempt to stave off the disintegration of the Spanish state rather than as a genuine attempt to promote autonomy and cultural difference. Federalism or other highly devolved structures can be seen as political–territorial strategies implemented in an attempt to reconcile the competing centrifugal and centripetal forces to which states are subject, as suggested in Chapter 3. Franco's highly centralized state was envisaged as a means of suppressing separatist tendencies. The granting of significant autonomy can be seen as a different way of dealing with these forces, while at the same time retaining the regions as integral parts of Spain.

The United Kingdom is currently moving towards a form of regional auto-nomy with parliaments having recently come into being for Scotland and Wales following referenda held in 1997. In a slightly different context Northern Ireland has its own elected assembly, part of an overall 'peace strategy' designed to end years of militant political conflict (discussed in Chapter 5). While it can be argued that these represent genuine attempts at enhancing regional democracy within the UK, critics might argue that they can be interpreted as efforts to maintain the union through offering limited devolved powers in order to stave off complete separation and, consequently, the dissolution of the existing state (Bogdanor 1999). Viewed in this way, such devolutionary strategies might be seen as an attempt at maintaining state hegemony rather than a weakening of it. (As a side issue it is worth noting that the construction of electoral boundaries may occasionally reflect a very overt attempt to retain power – see Box 7.2.)

Box 7.2   Constituency boundaries and gerrymandering

Within the formal political system another version of territorialization is utilized. This is the construction of geographically based political units for electoral purposes. Generally these are known as constituencies; areas in which prospective members of parliament stand. While constituencies may not have the same resonance as local authority boundaries and the like (in some instances, such as the Republic of Ireland, they may have more significance owing to a history of clientelism), they do provide an example of a territorial strategy in the realm of formal politics. Politicians may behave in a very clientelist way whereby they perform 'favours' for constituents in order to ensure re-election. They may also be keen to lobby on behalf of developments viewed favourably by the majority of their constituents. This might be reflected in support of particular proposals aimed at job creation in certain areas, or it might be reflected in opposition to schemes seen as detrimental to the local environment. This clientelist approach to politics can apply at all levels within the electoral hierarchy from MPs down to parish councillors and it highlights again the connections between place and politics and how territorial constructions can be utilized in order to bring about a desired outcome (whether from the point of view of constituents or the politician concerned).

Ostensibly constituency boundaries are meant to reflect objective criteria related to population size and demographic change. However, the construction of constituency boundaries is not necessarily a neutral process. It can be manipulated in ways which work to the advantage of particular parties or to the disadvantage of others. Even when the process is handled in an impartial manner there may still be 'accidental' consequences which could work to the advantage of one party or grouping. Gerrymandering refers to the deliberate manipulation of electoral boundaries so as to gain an advantage through a malapportionment of voters between constituencies. It usually involves the construction of electoral boundaries such that the votes of the opposition are effectively wasted or the advantage accruing from them is minimized while the manipulators' votes are maximized. The term was coined following boundary alterations by Elbridge Gerry, the Republican governor of Massachusetts, in 1812 which worked to the benefit of his party. The resultant electoral district was shaped like a salamander, giving rise to the term gerrymander.

There are two main types of gerrymander. The first is the excess vote technique whereby a particular recognizable subgroup of the population is apportioned a small number of constituencies where their political representatives win but within the overall context their number of seats is smaller than that of the party or grouping supporting the gerrymander. A second type of gerrymander is referred to as the wasted vote technique. In this instance constituency boundaries are drawn in such a way that the votes for a particular group are dispersed through a number of constituencies with the result

(continued)

(continued)

Table 7.1 Example of gerrymandering

Two 'normal' constituencies

	A	B
Blue Party	45	30
Red Party	40	60

Number of seats: 1 Blue, 1 Red

Four gerrymandered constituencies

	A	B	C	D
Blue Party	45	15	10	5
Red Party	40	10	9	41

Result: 3 Blue, 1 Red

that few, if any, of their preferred representatives are elected. Under these circumstances a party may win more votes but fewer seats (Table 7.1). Gerrymandering represents a deliberate manipulation of space in order to gain a political advantage.

In short, while devolution, federalism and related strategies may well generate more efficient service delivery systems, and while they may enhance local democracy (although neither of these is a necessary outcome), they are primarily forms of state territorial management. In order to maintain its territorial hegemony, the state has two main methods. The first is a policy of assimilation, the other is one of devolution. The assimilation argument is that all regions should be treated more or less equally with no effective concessions to local sensitivities (as in Franco's Spain). The devolutionist pathway is one where power is devolved to constituent territories. Ultimately both approaches are designed to prevent territorial secession and to maintain state control over all regions.

One qualification needs to be placed on the above arguments. The extent to which the state is ultimately able to control the pressures towards secession may vary. Clearly secessionist movements sometimes succeed, as with the fight for Eritrean independence from Ethiopia, while in other instances protracted and expensive military conflicts take place, as in Northern Ireland. Recent moves towards devolution within the context of globalization suggest that some substate entities see their future both locally and globally; thus, Quebec separatists envisage autonomy alongside integration into a wider world, as do Scottish nationalists. Interestingly, the moves towards greater autonomy in Wales and Scotland have led to pressures to create regional assemblies within England. These moves stem from desires towards greater local democracy but are not secessionist in nature, not carrying the force of a nationalist movement behind them. Moves towards a 'Europe of the regions' represent, to some extent at least, a bypassing of state structures as regional units make their own connections beyond the confines of their own state.

Community-based territorialization

The use of a sense of local community and attachment to place has been instrumental in the evolution of another dimension to new forms of governance at the local level. This is the inculcation of community-based responses to particular social or economic problems such as crime, rural development and service provision. These might be seen to fit into the wider growth of social movements associated with such issues as the environment. These moves are associated with the espousal of more bottom-up people-centred attempts to develop local responses to particular problems. Traditionally, planning models have adopted a 'top-down' paradigm with solutions and programmes implemented by professionals in the planning and development arena. More recently, within Europe, the idea of allowing local people and organizations a say in devising solutions to their own problems has become the new mantra. This is by no means a new phenomenon. There is a long history of local mobilization, particularly in urban areas, with various forms of neighbourhood groupings and other territorially based interest groups or protest groups. However, it appears to have assumed greater importance in recent years. Increasingly, partnership arrangements between statutory agencies, the private and voluntary sectors are being promoted and encouraged. This can be seen to be related to ideas of active citizenship.

This approach has in the past been particularly emphasized in urban areas and has been associated with programmes and initiatives centred around residents' groups, neighbourhood organisations and the like, particularly in inner city areas. Current initiatives within the arena of rural development in Europe also reflect this ethos with an increasing espousal of partnership approaches to the problems seen to beset many rural localities. The current rhetoric of rural development plays heavily on the role of the local 'community'. These territorially defined communities are envisaged as playing an integral part in the process of initiating and managing projects in their own areas. The argument here is that:

> policies that are sensitive to local circumstances will not only be more effective in taking the uniqueness of local social structure, economy, environment and culture into account, but also, through the involvement of the local community, will be more likely to be successful in their implementation. Communities that have a say in the development of policies for their locality are much more likely to be enthusiastic about their implementation. (Curry 1993: 33)

It might be argued that this shift reflects wider notions of moving away from a modernist vision of planning to a more postmodernist approach emphasizing rural diversity and local differences. In this way, locally sensitive initiatives are espoused rather than developing cross-spatial blueprints. Of course it also reflects the political programme of minimizing direct state involvement.

These approaches have arisen for many reasons. From an anthropological perspective it might be argued that humans have an innate tendency to co-operate in ways which refute theories of individuality outlined in Chapter 2. Shared residency may be one factor, although clearly not the only one, giving rise to a sense of shared identity at a micro-scale. Current moves might be seen as an attempt to build upon this sense of co-operation in order to genuinely demo-cratize local developmental processes by enabling and empowering local people.

Despite the rhetoric, however, it might also be argued that there are deeper reasons for the growing popularity of such approaches. One argument is that they provide a cheap response to particular problems and that they are not a huge financial burden on the state. A more sophisticated argument relates to the idea that what is happening is the use of a territorial strategy to maintain the hegemony of the state in ways similar to those outlined earlier. Viewed in Gramscian terms the current emphasis on community might be seen as facil-itating a process of incorporation whereby local actors and groups are drawn into the wider administrative structures, thereby blunting their radicalism. They are being brought under control and, in so doing, the state can ensure its continued dominance. Through the conceding of some rights the greater political edifice remains unthreatened. Thus, local people are 'involved' but are not necessarily in control. White suggests that participation has become a ' "hurrah" word bringing a warm glow to its users and hearers' (1996: 7) but one which may mean little in practice. Rather than government by community, current developments may well be a form of 'government *through* community' (italics added), as suggested by Murdoch (1997). In any event community groups and local partnerships are becoming part of the machinery of the local state rather than existing outside it.

This emphasis on voluntary activity at the level of local communities has also been in evidence in activities such as crime prevention, where people's sense of identification with their own neighbourhood is the key agent. Schemes such as Neighbourhood Watch in the UK are based on volunteers 'policing' their own localities and generally 'keeping an eye out' for anything unusual (Figure 7.8). The use of signs, usually displayed on lampposts, defines the boundaries of each Neighbourhood Watch area and acts as a clear territorial marker for those entering these zones (Plate 7.2). As with the other examples used above, it might be argued that this represents a cheap form of policing. It also ensures that local community activists remain 'onside' and become part of the policing apparatus rather than remaining outside it (Yarwood and Edwards 1995).

Whatever the motivation, it is clear that self-help and voluntarist solutions and the employment of partnership arrangements may, however, only offer piecemeal solutions and may merely serve to involve people but not alter the structures creating problems in the first place. Be that as it may, the interest in, and apparent enthusiasm for, such arrangements further highlights the extent to which people are willing to mobilize around a territorial identity. Ray (1998) suggests that, within a rural context, such strategies rest on the creation and maintenance of a strong territorial identity. Yet again, territorial

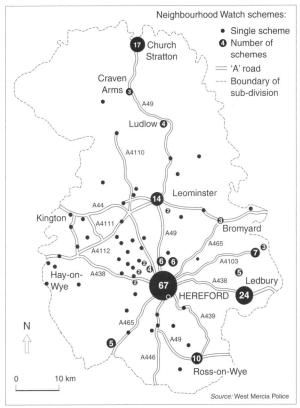

Figure 7.8 Neighbourhood Watch schemes, Herefordshire and South Shropshire, England. *Source*: Yarwood and Edwards, 1995

strategies are utilized and place-bound identities are seen as a key mechanism in fostering a particular spirit of innovation and co-operation. The distribution of these territories can be seen to exhibit a degree of spatial unevenness with such initiatives being more prevalent, or at least more successful, in areas with a longer history of co-operation or in various ways reflecting the levels of affluence of those involved (Storey 1999). To a considerable extent, the utilization of these territorial strategies is more a middle-class phenomenon, one supported by those with both the time and the money which allow them to become involved in these co-operative projects.

In a broader context these moves towards a version of community governance have a transnational dimension. Links between groups in different countries are mediated through a number of channels. Within the EU intergroup contact is facilitated through various networks, and programmes such as LEADER (concerned with rural development) are built around ideas of information exchange and cross-country co-operation. This is linked to other notions of a 'Europe of the Regions', emphasizing increasing use of contacts between regional governments and between community-based partnerships. While this does not seem likely, for the moment at least, to signify a complete bypassing

Plate 7.2   Neighbourhood Watch sign on a city street, Worcester, England.

of national governments, it does suggest a re-territorialization across international borders, linking into ideas suggestive of the diminishing importance of national boundaries as discussed in the previous chapter. It also highlights the interaction between the local and the global, whereby smaller territorial units are being brought into contact with localities far beyond their immediate day-to-day sphere of contact.

These ideas of community development reflect the shifting nature of local and regional government and planning referred to in the previous section. They fit into recent notions of the active citizen, popular with the 'new right', and which are seen to involve linking community (and, hence, people) to place. On the one hand they might be seen as locally sensitive attempts at deepening democracy; on the other they might merely be seen as a different means of retaining the state's territorial hegemony.

Administrative divisions and a sense of place

The above discussions have presented territorial strategies in broadly functional terms. The basic view propounded is that internal territorialization is a mechanism for the retention of political control. Despite this, however, the significance attaching to these territorial formations should be borne in mind. As was observed in relation to national identity and state formation, the erection of boundaries leads to the creation and sustaining of territorial identities. If particular territorial formations have a long history, or are assumed to have, then people may have a strong identity with their region or at least with some notion of their own locality. Territorial subdivisions within states may have long historical origins. Such is the case with counties in England. The reorganization of local government in 1973 led to the creation of many new counties which had no historical basis. In part, this reorganization was designed to create a number of metropolitan counties which would correspond more neatly to the reality of late twentieth-century Britain. For some people these entities were difficult to identify with in any meaningful way. Older deep-rooted senses of attachment to county, exemplified by local rivalries such as that between Lancashire and Yorkshire, proved deeply ingrained and have persisted despite the formal disappearance of some counties. Significantly, 'old' counties have recently been reinstated, in part perhaps a recognition of people's attachment to older territorial divisions and the failure of newer ones to gain broader acceptance in the public mind.

As suggested above, old county divisions never fully disappeared. County loyalty was affirmed and perpetuated through a variety of practices, through county cricket teams, local newspapers and other media which served to reinforce people's self-identification. These continued to operate and helped to maintain (and perhaps even strengthen) people's attachment to a particular construction of place and place boundaries. This adds further strength to the arguments in earlier chapters concerning both a sense of place and the extent to which boundaries, once created, become powerful elements in shaping identity. It can be concluded that administrative divisions may well have considerably more meaning in our everyday lives than simply lines on a map or providing a mechanism for determining service delivery. This is a further example of the way in which the state impinges on everyday existence and of how a sense of territorial identity may well reinforce state structures. Of course it is also the case that a sense of identification with these territorial divisions can be a useful mobilizing force for activities which can be seen as subversive. Various protest groups may organize themselves into county-based units to protest against what they see as undesirable activities on the part of the state. Once again, territory is being used as an organizing device in order to attain political ends.

Place and the private sector

Finally, it is worth noting that the utilization of a territorial strategy is not the preserve of government, whether in its centralized or highly devolved forms. The private sector may also employ territorial strategies in order to better manage their businesses; many companies operate on a regional basis. Obvious examples include electricity, water and gas companies in the United Kingdom. These all have their origins in the privatization of formerly nationalized industries. Moves towards privatization of public utilities in many countries (referred to above) mean that a range of previously publicly provided services are now provided by private companies, rather than state-run utilities. These companies tend to operate on a territorial basis. Within the UK, examples include Severn Trent Water in the English midlands and the electricity company Swalec in south Wales.

In addition to companies operating in specific regions, many organizations also use territorial strategies internally in order to effectively manage their activities. Territorial subdivisions are employed by a host of companies, such as those involved in life assurance, and are manifested through regional offices and regional managers and representatives with responsibility for specified geographical areas. In this way companies can more effectively respond to problems and demands which arise. Once again, territoriality is a useful organizational device and a means by which companies can improve their efficiency. The use of increasingly sophisticated market research techniques, utilizing postcode data and other geo-demographic information, has allowed companies to identify 'territorial markets'.

Companies may also utilize place knowledge in deciding where to locate. Hanson and Pratt (1995) report on a company in Worcester, Massachusetts, who located in a specific area because of the presence of a large Hispanic population which, the company felt, had the appropriate work ethic required by the employer. Ethnic stereotyping, combined with spatial concentration, entered into the industrial location decision.

Summary

This chapter has introduced just some of the more obvious forms of substate territorialization. It has provided an analysis of the key forms of substate governance as well as introducing key examples of private sector as well as quasi-state territorial strategies. It has also applied the theoretical insights introduced earlier in the book to an analysis of this territorialization and has suggested that the interaction between the central and local states is far from simple as modes of governance become increasingly complex. What is very apparent, however, is the importance of territorial arrangements in the day-to-day running of the state. Territorial strategies are used in order to manage the state and to ensure its reproduction. While these strategies can be seen in purely functionalist terms, it is important to bear in mind that the divisions used may help to foster or to reinforce a sense of identity; they are not merely functional containers.

Further reading

On federalism see the various chapters in G. Smith (ed.), *Federalism. The Multiethnic Challenge* (Longman, London, 1995). For a useful overview of dimensions of internal political territorialization in the UK, see J.B. Cullingworth and V. Nadin, *Town and Country Planning in the UK* (12th edition, Routledge, London, 1997). For more in-depth reviews of the nature of local government in Britain, see the Macmillan series 'Government Beyond the Centre', especially D. Wilson and C. Game, *Local Government in the United Kingdom* (1994) and C. Gray, *Government Beyond the Centre. Sub-National Politics in Britain* (1994). See also Liam Byrne, *Local Government Transformed* (Baseline Books, Manchester, 1996) and Allan Cochrane, *Whatever Happened to Local Government?* (Open University Press, Buckingham, 1993).

On the changing nature of service provision, see D.M. Hill, *Citizens and Cities. Urban Policy in the 1990s* (Harvester Wheatsheaf, Hemel Hempstead, 1994), while a useful collection of articles on community participation and empowerment appears in *Journal of Rural Studies* 1998, 14(1). On Neighbourhood Watch in rural areas, see the article by R. Yarwood and B. Edwards, 'Voluntary action in rural areas: the case of Neighbourhood Watch', *Journal of Rural Studies* 1995, 11(4): 447–59. On the spatial dimensions of apartheid see D.M Smith (ed.), *The Apartheid City and Beyond. Urbanization and Social Change in South Africa* (Routledge, London, 1992).

Chapter 8

Territory and locality

As suggested in Chapter 1, territoriality is a phenomenon not confined to the world of formalized politics. At more micro-levels territorial strategies may be used in attempts either to attain or to retain power or to bring about certain changes. This chapter focuses on issues related to social inequalities and indicates how, firstly, these inequalities are mapped onto space and, secondly, how social life affects, and is affected by, territorial formations. In the main, these are territorial phenomena which are less obvious and which may be more difficult to detect and observe, relative to those discussed in earlier chapters. Although they may appear more nebulous, this does not make them any less real, nor does it render them essentially different from the more readily defined territoriality identified in previous chapters. While the processes occurring may be different they can still be seen as the spatial reflection of power and each of the examples chosen reflects the interconnections between geographical space and everyday life.

Thus far in the book, discussions of territoriality have centred on what can be seen as the public arena of politics in its various forms. Feminist geographers in particular have been instrumental in focusing critical attention on divisions between the public and private domain and its spatial corollary of a division between what is seen as public space and private space. It is argued that issues of gender interact with space and place in various ways. For example, some spaces, streets or parks may be deemed unsafe for women, particularly lone women. In this way, issues of gender are mapped onto space. Within this chapter reference is made both to the public domain, mainly divisions within urban space, and to the private domain, mainly focusing on domestic space as seen in the household. There is a focus on issues pertaining to ethnicity; class; and gender and sexuality.

Underpinning the examples used here, in part at least, is a sense in which particular spaces become characterized as the preserve of a specific set of individuals. Boundaries are constructed and contested in peoples' everyday lives. These boundaries reflect broader issues related to power relationships, relations based on class, ethnicity, gender or sexuality which give rise to significant social inequalities. The construction of these boundaries engenders a sense of people being in their 'proper' place or 'out of place'. While many people do not necessarily freely choose their 'place', they may, nevertheless, identify with their immediate neighbourhood or locality. This sense of identity can in turn be converted into forms of action aimed at obtaining particular outcomes. The formation of community or residence groups reflects feelings of belonging or attachment to a particular place. In line with earlier discussions, it follows that notions of territory are connected with ideas of social power. The claiming of space is a political act whether it occurs in the public or private arena.

This chapter explores micro-level territory and territorial behaviour under four key headings. Firstly, it explores the development of what can be seen as racialized spaces; secondly, it considers the organization of space along other social or cultural fractures, with a key focus on territorial divisions based around class and religion; thirdly, it examines the issues of gender and sexuality and their territorial expression. Finally it briefly examines issues relating to territorial practices in the home and the workplace. In line with earlier discussions, the object is to examine the ways in which social phenomena manifest themselves territorially and how territorially based strategies may be utilized in the pursuit of certain objectives. While the various topics discussed in this chapter are examined in relatively discrete sections, it should be abundantly clear that many of the issues raised are interrelated. As suggested earlier in discussions of national identity, people have multiple identities. Gender, sexuality and ethnicity cross-cut each other in a system of overlapping identities. Finally, a cautionary note on the social categories used in this chapter. Ethnic groups, religions, classes are not immutable, they are social constructs. As such we need to be aware of the dangers of seeing them as rigidly defined.

Racialized spaces

Urban areas exhibit various forms of spatial differentiation, in terms of building type, plan and layout, land use (industrial, residential, etc.). Socially, cities and towns are also characterized by various forms of residential segregation. As Carter has observed 'the concentration, segregation and isolation of population groups within the city has been characteristic since earliest times' (1995: 261). The study of residential differentiation in urban areas has proceeded through many stages since the early days of the Chicago school of urban ecology and ideas of 'natural areas' within cities. The energies of many urban geographers have been taken up in the use of social area analysis and factorial ecology in order to rigidly classify areas within cities based on a range of social and economic criteria (Figure 8.1). There is a wide variety of reasons underlying this well-established trend of spatial segregation. It is a feature very often associated with ethnicity; however, segregation along class lines is an equally well-established phenomenon. Certain areas may become the effective preserve of one ethnic group or of one class. In this way we can speak of racialized spaces or of working-class or middle-class areas.

Race is seen by many social scientists as a social construction rather than a biological reality. While race can be questioned as a dubious form of social classification, there is no doubt that racism or 'race thinking' is a very real social phenomenon. Jackson defines racism as

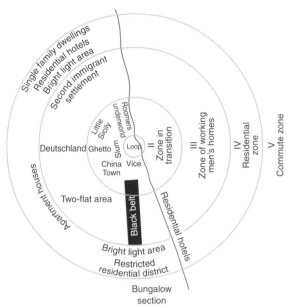

Figure 8.1 Burgess model of urban land use: Chicago in the 1920s.
Source: Knox, 1995

the assumption, consciously or unconsciously held, that people can be divided into a distinct number of discrete 'races' according to physical, biological criteria and that systematic social differences automatically and inevitably follow the same lines of physical differentiation. (1989: 132–3)

Although there may be little basis for a racial categorization of people, that does not prevent some people from behaving as if there was. Race and racism can obviously be expressed in many ways, not the least of which is geographical. Particular spaces may come to be seen as the preserve of a particular 'racial' group. In the case of a group deemed to be inferior or subordinate, their presence in particular areas may often be seen as undesirable, as a source of contamination and as something to be outlawed. Such non-dominant groups are seen as the 'other'; they are deemed to be different from that which is considered the norm. The classic example of racialized space was that devised under the apartheid system in South Africa; a territorial system which enhanced political, economic and social power of whites over blacks (see Chapter 7). Although apartheid no longer officially exists, its legacy in South Africa means that a division of space into black and white 'territories' continues based on the racial and class lines reinforced during the apartheid era and which will take many years to dissolve. Many black people cannot afford to move out of the areas in which they find themselves. 'Squatter' settlements composed of black residents still evoke hostile reactions from many whites. These settlements arose on the fringes of 'white' areas. This very designation symbolizes in its purest form the racialization of territory. Discourses linked to ideas of purification and defilement continue to surround debates over squatter settlements;

black people are seen to be out of place in certain locations. The idea that they should stay in their areas is a view which has not yet disappeared and is unlikely to for some time to come (Dixon, Reicher and Foster 1997). Racial segregation and the creation of bounded spaces was the spatial expression of white power. Non-white people were 'placed' in locations not of their own choosing in order to enhance minority white power. In this way, there was a legal transposition of inequality onto geographical space.

Despite the overt nature of South Africa's racial segregation, it should not be thought that this is the only form of territoriality based on race or ethnicity. In a town in the Czech Republic in the late 1990s a wall was constructed around an area of housing occupied by gypsies, a group experiencing considerable discrimination particularly in eastern Europe (Grant 1998). Gypsies are seen as an undesirable 'other' as a consequence of which they are effectively de-territorialized; they do not belong anywhere (Fonseca 1996). The furore over the recent arrival in Britain of a number of gypsies from the Czech and Slovak Republics emphasizes this sense of a group of people seen not to belong in any place; a people without territory.

Another example of this sense of an 'other', a group seen to be 'alien' to a particular place, is provided by May (1996). Referring to Stoke Newington in London, he demonstrates that a variety of senses of Englishness led to the view that the area's purity is somehow reduced owing to the presence of an immigrant population, mainly Kurdish in origin. This reflects a view that some categories of people, defined in terms of 'race', should not be allowed into certain places; they are seen as a source of contamination altering the true essence of a mythically 'pure' place. A variety of factors serve to support this ideology: fear of the unknown, a deeply felt hostility to the 'other', to those seen as different, economic fears associated with the belief that ethnic minorities are taking 'our' jobs, and so on. Such perspectives privilege 'us' over 'them' and contribute to the idea that these 'others' should be confined to 'their' places and not enter 'ours'. Short has argued that 'the greater the degree of otherness of an ethnic group the more spatially segregated they are' (1996: 222). In rural areas members of ethnic minority groups may be seen to be 'out of place'. The photographer Ingrid Pollard has spoken of feelings of unease in the countryside and her photographs conjure up ideas suggestive of black people as not belonging there (Kinsman 1995; Plate 8.1).

Ghettoization

In many urban areas in Europe and north America, particular neighbourhoods exhibit concentrations of specific ethnic groups. Toxteth in Liverpool, St. Paul's in Bristol, Brixton in London are seen as 'black' areas, Balsall Heath in Birmingham as an Asian area, Kilburn in London as an Irish area, and so on. In popular parlance the term 'ghetto' is often used to describe such areas. However, the term more precisely refers to an area in which the population is almost exclusively drawn from a particular ethnic or 'racial' background (Peach 1996).

Plate 8.1   Ingrid Pollard.

Within the urban geography literature, there is considerable discussion of the reasons underlying such concentrations. Following Boal (1978), a number of influencing factors can be identified (Table 8.1). In brief there is a combination of 'positive' and 'negative' factors; for some there are attractions such as 'being among one's own', while others may feel driven to seek sanctuary from a racist, hostile society. However, this categorization might tend to suggest that ghetto residents have freely chosen to be there. In reality, the degree of choice available to many ethnic minority people is extremely limited. The idea that people may choose to cluster is to ignore the fact that quite often no viable alternatives are available. The wider structural constraints within which

Table 8.1   Factors influencing ghetto formation

- *Defence* against attack by the majority group
- *Avoidance* of the dominant culture in order for the minority to assert its own identity
- *Preservation* of cultural norms and heritage, which is easier when one is part of a larger group rather than an individual or lone household
- *Resistance* to threat or assimilation can be enhanced through the use of a territorial strategy; political representation may be obtainable as a consequence of significant numbers of the same ethnic group residing in the same locality

Source: Boal, 1978

minority groups live militate strongly against liberal ideas of free choice. In many ways, it is argued, ethnic minority groups become trapped in particular locations as a consequence of their peripheral position within the labour market which renders them unable to afford owner-occupation. Allied to the actions of housing 'agents' (such as local authorities, estate agents) which may tend to 'push' people towards certain areas, or discourage particular groups from moving into certain locations, such groups may become trapped in inner city public housing. To a considerable extent it could be said that members of certain ethnic groups face a highly constrained choice in terms of where to live. In any event, a group which is seen to be distinctive and which is in some way (quite often in terms of skin colour) seen as the 'other', distinct from the majority population, is pushed into its own residential area as a consequence of the processes of discrimination and exclusion operating in wider society. Thus, it might be suggested that 'external' factors are ultimately of greater explanatory power than Boal's internal ones. Discriminatory ideologies of race work to exclude people from particular milieux, thereby translating social exclusion into geographical exclusion.

It is not just the case that ethnic minority groups cluster in certain ways in response to prevailing economic conditions and as a reflection of various discriminatory practices (although this is undoubtedly the case). It should also be borne in mind that this clustering itself further contributes to future rounds of marginalization and exclusion. As Susan Smith eloquently puts it, 'what was a spatial reflection of economic and social marginality becomes a spatial constraint on economic advance and social mobility' (1999: 18). This further demonstrates the idea that territorial practices serve to reproduce particular social outcomes in a recursive relationship whereby society does not simply impact on space, but spatial arrangements in turn impact on society.

In considering the ghetto, there is a tendency to see it in quite negative terms. The ghetto is a territorial entity and it is one which evokes many negative connotations; the term is often seen as synonymous with 'slum', the juxtaposition tending to lead to a stigmatizing of its residents. In some popular interpretations the ghetto is seen as an area populated by a group of people with dangerous moral standards and a general disregard for wider society. In this way the ghetto is seen almost as a separate territory, virtually a *bantustan*, separate from the rest of the city and occupied by a threatening 'other' who need to be kept apart from the rest of society (Wacquant 1994). In such ways the hegemony of the dominant group is maintained and the 'other' (in this case, ethnic minorities) remains geographically, as well as socially and economically, marginalized.

Even leaving aside overtly prejudiced readings of the ghetto, there are those who see it solely in terms of poverty and destitution and who regard its evolution as entirely regrettable. There has been a distinctive literature arguing that the ghetto is inhabited by a relatively poor 'underclass', a term suggestive of moral and social differences distinguishing its members from mainstream society (Banfield 1958; Lewis 1966; Hill 1994). Ghettos and other inner city areas are often seen as the home of people locked into a cycle of deprivation and disadvantage (Rutter and Madge 1976). They are portrayed as sites of violence in which street gangs hold sway (see Box 8.1). In this way the ghetto is portrayed as a problem rather than a place whose residents experience problems.

The ghetto is a territorial manifestation of social inequality and reflects the uneven distribution of power in society. Of course, like any other territorial entity, the ghetto can be used as a means of mobilizing residents, of providing the territorial frame within which people can operate with a view to improving their own conditions. Successful residents' groups often emerge which may become sufficiently well organized to be able to engage in lobbying, in forms of self-help, etc. Such forms of community action, touched on in the previous chapter, may well make a positive difference to the lives of ghetto residents. In this way, it can be argued that the ghetto, just as with any other territorial formation, can take on its own sense of identity and can become a mechanism for the expression of group interests. Normal systems of political power may be undermined and partially subverted through this process. This does not of course obscure the reality that its occupants remain largely on the fringes of mainstream society.

Box 8.1   Urban gangs

The activities of urban street gangs have received considerable media attention in recent years, particularly in the US. These gangs engage in occasionally quite violent conflict. The behaviour of these gangs has a strong territorial component. Their *raison d'être* is the assertion of control over their 'turf'. Rival gangs are not welcome on their patch. This territorial behaviour might be seen as reflecting some sort of innate territoriality, but it can also be interpreted as a consequence of the marginalization of many poorer young people in impoverished urban areas. Territoriality may be a means of expressing power using the only resource available to them, the streets and neighbourhoods in which they live. These gangs may be linked to criminal behaviour and may be involved in controlling illegal activities in their patch, mirroring the behaviour of 'older' criminal gangs who also display a territorially based organizational structure.

Street gangs tend to lay down territorial markers to indicate to others their 'ownership' of particular places. The use of graffiti on walls, bridges and buildings is one very visual method of claiming space. Markers are quite literally placed on the landscape to signal control of territory or 'turf ownership'. Work in Philadelphia indicated that graffiti became denser closer to the core of that gang's territory. Ley and Cybriwsky (1974) were able to demarcate reasonably accurately the spatial extent of gang control in the city. In this way aspects of popular culture are translated into a territorial frame. The claiming of space is what is important. It may be a means by which marginalized youths make their claim to existence; through planting their mark on territory, that territory becomes theirs.

Areas with high concentrations of ethnic minority groups can have much more positive connotations – areas such as the so-called 'Banglatown', centred on Brick Lane in London, which is predominantly inhabited by Bengalis. As an area with a long history of Jewish and Irish immigration it can be seen as epitomizing a certain cosmopolitanism which can act in a very positive way for local residents (Dwyer 1999). However, it can also lead to a targeting of such areas by racist groups. In 1999 a nail bomb attack was carried out in Brick Lane; an apparent assault on the area's non-white ethnic groups. Thus, the promoting of a multicultural space (a 'diasporic space', as Dwyer refers to it) is met with resistance from those who wish to see such spaces remain ethnically 'pure'.

While ghetto areas in some ways reflect the dominant power relations within society, Bozzoli (1999) shows how a South African township in the 1980s, a space designed and built by the dominant power to house subordinate groups, became transformed into a place of resistance. Spatial arrangements engender specific social relations as a consequence of human interaction. Spaces can take on different meanings to those originally intended. As she puts it, 'space develops over time a "hidden transcript" of its meaning to those who inhabit it, different from the "public transcript" of its meaning for those who rule' (1999: 6). Residents of Alexandra township in Johannesburg worked to make the place ungovernable through a combination of tactics including an action committee and through symbolic territorial actions such as patrolling the streets and the actual renaming of streets to reflect their struggle (for example, one street was renamed ANC Street). The space that was Alexandra was being re-shaped into something different from that originally intended. It became a space of resistance rather than merely a space of oppression:

> Sometimes the places in which the poor and oppressed live start off as bounded and prison-like stalags. But there are times when rebellious inhabitants seek to transform the stalag into a space of their own upon which their meanings are imprinted and whose boundaries become the defiant barricades which keep the authorities out, rather than the symbolic walls which keep the persecuted in. (Bozzoli 1999: 40)

This section has demonstrated the ways in which racist and exclusionary ideologies are transposed onto space. It has also indicated ways in which those racist constructions are opposed. Just as particular power relations are refracted through a territorial frame, so those relations are contested through territorial strategies. The spaces to which people are consigned may provide the means through which they contest their marginalization.

Social and cultural divisions

Segregation by class

The example of racialized space provided above overlaps with other forms of segregation, essentially those based on social status. Most cities have distinct residential neighbourhoods, colloquially defined as 'rich' or 'poor', 'working class' or 'middle class'. If the apartheid system created formalized 'racial' spaces, then less formalized but equally effective segregation – often overlapping with race – has occurred elsewhere. Segregation along class lines is effected through various agencies and through the operations of the housing market, effectively determining who can afford to live where. This might be said to be given its most formalized vestige in cities in the United States through the process of municipal incorporation (Figure 8.2). Under this system, better-off territorially defined urban areas can effectively secede from the larger city of which they are a part. In doing so they enjoy a degree of fiscal autonomy which means that residents do not have to support services, such as public transport, for poorer areas outside their own municipality. In this way, the area opts out of the broader urban environment and so sheds itself of any sense of collective responsibility for the poor. The incorporated municipality can decide on certain local regulations such as excluding industrial developments. It also has the

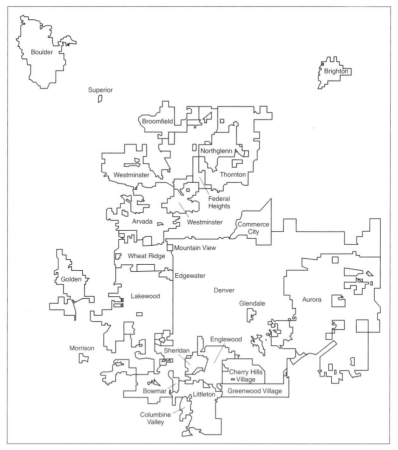

Figure 8.2   Separate municipalities, Denver urban area, Colorado.
Source: Johnston, 1982

power to enforce minimum lot sizes and prohibit mobile homes, and so on. In this way it can effectively exclude poorer residents. While this is essentially a form of class-based territorial exclusion, in many instances it has racial implications. Overt racial discrimination may be illegal but black people, or other non-white groupings, may well be excluded through the implementation of regulations excluding the poor who, in most US cities, are disproportionately black or Hispanic. This is another clear example of the political manipulation of space. Power is being mediated through a territorial process which may have serious racial as well as class connotations.

A more overt example of the way in which the rich can exert territorial power over the poor is provided by Mike Davis in his book *City of Quartz* (1990). He refers to parts of Los Angeles where security guards patrol the perimeter of walled residential zones in an effort to exclude what are seen as 'undesirables'. This is a classic example of how poorer people are excluded through this privatization of urban space. Davis refers to this as 'Fortress LA' and it is a phenomenon repeated in many north American cities where residents can effectively govern their own territory resulting in the exclusion of those they deem undesirable or unsuitable. Versions of these 'gated' societies also exist within the UK with private security firms patrolling more exclusive urban areas in order to exclude those seen as undesirable, thereby maintaining the 'undefiled' nature of the neighbourhood.

Similar processes of exclusion, although perhaps less overt, could be said to operate in the construction of places such as London's Docklands. Indeed, even the notion of constructing new housing developments aimed at specific sectors of the population, in this case a young upwardly mobile middle class, serves to reinforce economic divisions within society and perpetuate the idea that some households do not belong in particular places. As Short (1989) suggests, the built environment reflects the perceived needs of different household types and social categories. Gentrification and the construction of 'yuppie' apartments in inner urban areas reflect broader socio-economic processes and the resultant residential territorialization can be seen as an expression of the financial power of home-owners and the power of finance capital. In this way places like London's Docklands are transformed from manufacturing and working class residential areas into service sector (particularly financial services) zones with a resident middle-class population. This, as Short points out, reflects more than a simple change in land use; it also reflects changes in the meaning of place. Docklands, like other 'regenerated' urban zones, has been transformed into a different place, with quite a different symbolic meaning.

Associated with these transformations, it can be argued that the very design of buildings reinforces notions of territorial separation. Many middle-class urban residential blocks clearly exclude those deemed undesirable. The placement of gates and intercom systems further enhances this system of exclusion. Defensible space is seen as important. These territorial strategies work in ways which ensure a particular residential mix and may serve to link both racial and class divisions. Conversely, many working-class housing estates, seen as being inhabited by an underclass, are often seen as 'no-go' areas; as territories out of bounds. Many 'deprived' estates acquire a negative reputation. Such stigmatizing of place in itself becomes part of the problem, reinforcing class divisions and reproducing various forms of social exclusion.

Religious segregation

Just as ghettos arise owing to forms of racial segregation, religious differences can also result in territorial separation within urban areas. One of the best-known examples is that which occurs in Northern Ireland. The broader political situation here has been dealt with in previous chapters but, as in the case of South Africa, larger-scale territorial disputes also have micro-scale impacts at the level of individual neighbourhoods or even streets. In urban areas in Northern Ireland, there are Catholic residential districts and Protestant districts, linked to the region's long-standing political problems. In terms of public housing there are effectively two parallel sectors, thereby ensuring that segregation continues. Indeed, discrimination against Catholics in the allocation of public housing was one of the factors giving rise to civil rights agitation in the 1960s, thereby contributing to the on-going conflict. While overt discrimination of this sort may now be less frequent, the patterns continue for the reasons suggested earlier in relation to ghetto formation. Essentially, cities such as Belfast have what are, effectively, Protestant ghettos and Catholic ghettos (Figure 8.3). However, it should be borne in mind that this ethnic reading of the conflict can be slightly misleading; not all Catholics are nationalist and similarly

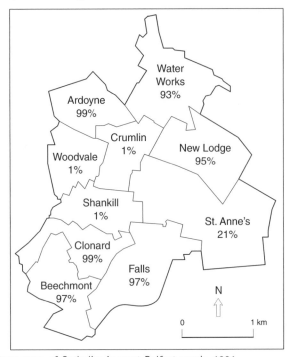

Figure 8.3   Percentage of Catholics in west Belfast wards, 1991.
Source: Kitchin and Tait, 2000

Plate 8.2 'Peace wall' dividing Catholic and Protestant areas in west Belfast.

not all Protestants are unionist. Nevertheless, it is apparent that there is considerable religious segregation even if this cannot, and should not, be seen to correspond completely to peoples' wider political beliefs. Once again, this spatial segregation reflects power imbalances and highlights the way in which political relationships are mapped onto space.

In Belfast, where the so-called Shankill–Falls divide separates the two groups in the west of the city, a very high degree of segregation occurs, highlighted by the inappropriately named 'peace line' – sections of intermittent wall literally dividing roads in the area and designed to prevent confrontations (Plate 8.2). The extent of segregation and the lack of intergroup mixing were highlighted in one study which indicated the tendency of people to use shops, and even bus stops, in their own area, even when one on the 'other side' was geographically closer (Boal 1969). Parts of the city are seen as out of bounds to one side because it is the other's territory. Once again, this highlights the territorial nature of political conflict and emphasizes the manner in which territory very easily comes to be seen in terms of 'ours' and 'theirs'. The territoriality of Belfast is such that a person's address is quite likely to reflect their place in a religious as well as a merely geographical sense. Parades by the Protestant Orange Order can be seen as attempts to cling to territory, if only metaphorically.

This intense spatial divide is highlighted in the experiences of people growing up in Belfast. In a recently published memoir, Ciaran Carson, a poet from a Catholic background, quotes from the writer Robert Harbinson on his experience of this territoriality. From Harbinson's Protestant perspective certain places were out of bounds (note: 'Micks' and 'Mickeys' are derogatory terms for Catholics):

God ordained that even the Bog Meadows should end and had set a great hill at their limit, which we called the Mickeys' Mountain . . . In terms of miles the mountain was not far, and I always longed to explore it . . . But the mountain was inaccessible because to reach it we had to cross territory held by the Mickeys. Being children of the staunch Protestant quarter, to go near the Catholic idolators was more than we dared, for fear of having one of our members cut off. (Harbinson, cited in Carson 1998: 86)

Juxtaposing Harbinson's childhood memories with his own view from the other side, Carson reminisces about crossing 'the slippery stepping-stones across the Blackstaff into the margins of enemy territory, which we approached with the same trepidation felt by Robert Harbinson, coming from the other side' (1998: 86). Both writers are expressing their sense of the city and the significance of the political and religious divide within it which had a direct territorial impact on their childhoods (and those of many others).

Just as international borders have flags and other territorial markers to indicate their location, so also these segregated residential areas come complete with their own sets of boundary signifiers. Red, white and blue kerbstones indicate loyalist areas, while Irish flags and a variety of republican wall murals are highly visible symbols in nationalist areas (see Box 8.2). Similarly, graffiti

---

Box 8.2   Wall murals in Northern Ireland

The drawing of wall murals in Northern Ireland demonstrates the use of historical images in order to make contemporary political statements. It represents a micro-scale version of this larger conflict. While utilizing images of individuals or of historic events the placing of these murals in public spaces, gable walls of houses or on other spaces which are easily seen serves a number of territorial functions. It uses a particular space to broadcast a political message, as with political graffiti. While a prime purpose of murals can be seen as sending a message to the muralists' own community and performing a politicizing function, they can also be seen as territorial markers. They send out a message that this space belongs to this tradition. Indeed some of these murals occur on the boundary between nationalist and loyalist areas in Belfast (Plates 8.3 and 8.4).

There is a longer history of mural painting in loyalist areas. Indeed, it has been argued that such painting was instrumental in the 'construction' of 'Protestant areas'. In recent years, it has been observed that Republican murals tend to adopt a more cultural form while Loyalist murals cling to a more militaristic set of images of gunmen, weapons and a continued assertion of Ulster's defence (Rolston, 1997–98). Equally, it has been suggested that loyalist murals tend to be somewhat insular while republican murals are more likely to reflect wider international connections, to places such as Cuba, South Africa and so on. These republican markers are occasionally written in the Irish language, an act itself signifying political allegiance to Ireland rather than Britain, and pointing to a cultural independence. The erection of 'unofficial' Irish language street signs could be seen as a similar territorial strategy. Murals may refer to specific

(continued)

(continued)

Plate 8.3 Republican wall mural in Northern Ireland advocating withdrawal of British troops. Photo courtesy of Liam O'Hare.

Plate 8.4 Loyalist wall mural in Northern Ireland in support of the Ulster Volunteer Force. Photo courtesy of Liam O'Hare.

events or places, some often include maps, whether of all Ireland (republican) or of Ulster only (loyalist), while many commemorate events which occurred on the street or in near proximity to the actual site.

says 'Prods keep out', 'no pope here', 'Ulster is British' or 'tiocfaidh ar lá' ('our day will come', a common republican slogan). These markers serve both to reassure residents and to send out a clear message to the 'other side' that their presence is not welcome. For people who find themselves living in – or even passing through – the 'wrong' area, the consequences can be fatal. These areas are quite often patrolled, to some extent controlled, by members of paramilitary organizations. Certainly in the past the Provisional IRA has seen itself as a defender of Catholic areas of Belfast while organizations such as the UDA (Ulster Defence Association) and UVF (Ulster Volunteer Force) have portrayed themselves in a similar light in relation to Protestant areas. Once again territoriality allows a mobilizing process to take place.

Territory, gender and sexual orientation

Gendered spaces

The growth of the so-called 'women's movement' from the 1960s onwards, built on earlier attempts to achieve equality for women, has been influential in achieving recognition for the unequal status of women in all dimensions of life; in the home, in the workplace, in the broader political arena. Feminist writers and activists have been instrumental in attempting to explain how patriarchal systems of power have tended to reinforce male dominance and how women have been consistently marginalized throughout history. Feminist geographers have highlighted the 'geography' of discrimination against women, particularly in drawing attention to the manner in which space and place are heavily gendered. More specifically, there has been a focus on the way in which patriarchal systems of power have led to a division between public and private space and have resulted in social practices which see certain activities and certain spaces as male preserves. In its most simple form this is reflected in the idea that 'a woman's place is in the home'. The implications of gender are seen to be as important as other social and economic factors in the structuring of spaces and places.

Underpinning this are ideas which distinguish between sex as a biological reality and gender which refers to the socially constructed nature of both male and female identities (Women and Geography Study Group of the Institute of British Geographers 1984). This moves us towards the importance of nurture and away from deterministic notions of human identity and behaviour, as suggested in Chapter 2. Thus, as with biological ideas of race, the argument is that as individuals we are not genetically predetermined to be more suited to some roles rather than to others.

The geographer Doreen Massey (1994) recounts how, as a teenager, she used to get a bus into central Manchester on Saturdays and how she was struck by the fact that football pitches were spaces inhabited totally by males. In this way, she argues, there is a clear gendering of space whereby some spaces come to be seen as the preserve of one or other gender. To take Massey's example one step further, it can be argued that professional football stadia were, at least until quite recently, almost exclusively male preserves. With a resurgence of interest in football in the 1990s this phenomenon has altered slightly with an increase in the numbers of female supporters and clubs encouraging family attendance. The relative growth in popularity in women's football is another dimension to this. Nevertheless, this division reflects ideas about separate leisure activities and, hence, separate spaces for women. Until comparatively recently pubs were a male preserve; a space in which women were, if not formally excluded (although in many instances they were), then made to feel unwelcome. In an era of a somewhat greater acceptance of the idea of equality of the sexes (although far from universal and, in many cases, more token than real), such exclusionary tactics may be less overt, although many would argue that they still persist, albeit in more subtle forms. Again, viewed from a feminist geographic perspective, such social practices are built on ideas of what is or is not acceptable behaviour for women to engage in (built on socially or culturally constructed notions of masculinity and femininity); in other words, the everyday reproduction of patriarchy. All of this implies that spatial or territorial strategies can be used in order to retain social control over women. Social processes reproduce attitudes which maintain that women adopt specific roles or modes of behaviour and that these roles and behaviours be acted out in specific spaces, e.g. 'home-making' and child-rearing in domestic space. Behaviour is thus not just gendered but also spatialized.

One reason for the absence of women in particular places is overt discrimination or active discouragement in the sense of certain activities or pursuits not being deemed suitable for women. Women who transgress these boundaries are often portrayed in a negative light; an idea reflective of notions of 'good' and 'bad' women. Women out alone at night might be seen as not conforming to what is expected of them. Despite the persistence of ideas such as these, it might be argued that there is now a greater acceptance of the idea of equality of the sexes (albeit with the emergence of 'new laddism' some might question whether this is the case) but there remains the problem of suitable role models for women and the achievement of a certain critical mass. Using the example of football alluded to above, it might be argued that until a significant number of women start going to football matches then it is harder for individual women to assert their right to be in such places. However, such an idea, fitting in with a liberal feminist view of equality, might be said merely to reflect women conforming to a masculine world by buying into male forms of behaviour. A more radical feminist perspective would argue for a celebration of women's difference and equality of treatment for women's leisure activities rather than an adoption of male norms.

A crucial aspect of the relationship between women and place centres on the perception of some specific places as 'unsafe'. Many women do not feel safe in certain public places, most notably darkened streets. These streets and alleyways and similar places become areas in which women feel threatened because of a fear of attack by men. As Valentine (1989) suggests, women transfer a fear of men into a fear of certain spaces. Women are conditioned into seeing male-dominated space as a threat, especially at night. Women thus engage in a process of self-policing whereby they reinforce patriarchy by accepting its premises – that they should avoid particular spaces. This has profound implications for the manner is which women negotiate their way through urban space (Fell 1991).

If women are seen as 'out of place' under certain circumstances then it follows that they 'belong' in other spaces. If public space is seen as being more properly a male environment, then women are seen to belong in domestic space. This is seen as their territory, corresponding to their presumed affinity with home-making and child-rearing. This division of labour confines women to the private realm leaving men to inhabit (much of) the public domain. This view of women as playing a subordinate role has in the past been reflected in discriminatory attitudes and practices, particularly in relation to women in the paid workforce with active discouragement through lower wages, if not actual exclusion, from many jobs. These views are predicated on the undesirability of women going out to work. Massey (1994) has argued that opposition to women going out to work was greater than opposition to women working *per se*. It can be argued that this ascribing of women's roles, through delimiting the spaces in which women were allowed to appear, is another spatial expression of power. In other words the confining of women to domestic space, and their exclusion from male territories, was the key element in male control. McDowell (1983) has referred to urban areas as containing masculine centres of production and female centres of reproduction. With increasing female participation in the workforce and a raft of anti-discriminatory legislation in many countries, such a generalization may appear to have lost some of its validity. The history of women's involvement in the paid labour force is now more differentiated and, indeed, displays considerable spatial differences (McDowell and Massey 1984). Nevertheless, the division between a (largely) male public sphere and a (largely) female private sphere still has considerable resonance.

Even when women enter the workforce, they may still encounter territorial divisions in the workplace. Thus, Spain (1992) alludes to the 'closed door' jobs of managers (mainly men) and the 'open floor' jobs of manual workers (who may be predominantly women in certain countries, regions or sectors). Again, this can be interpreted as a form of control by which workers' actions are constantly subject to monitoring. This has echoes of Foucault's idea of 'panopticism' whereby the spatial layouts of buildings and institutions facilitated control over the inmates (Foucault 1977; Philo 1989). Many employers may even locate in particular places in order to take advantage of what they see as an appropriate workforce whether this be seen in terms of gender, ethnicity or prevailing wage levels or skill levels. Hanson and Pratt (1995) show how some firms located in particular parts of Worcester, Massachusetts, in order to avail of what they saw as desirable ethnic or gender characteristics (see Chapter 7). The latter was related to jobs which the employer saw as suitable for women such as sewing, knitting, etc. In this way social segregation is maintained through industrial location decisions.

Domestic space

If women are seen as 'belonging' in the home, it is not the case that all parts of the home are equally theirs. Even within the home, territorial divisions take place; the most obvious being the notion of the kitchen as a female preserve. The neglect of such territorial behaviour in all its manifestations is commented on by Sibley:

> In geography, interest in residential patterns wanes at the garden gate, as if the private province of the home, as distinct from the larger public spaces constituting residential areas, were beyond the scope of a subject concerned with maps of places. (Sibley 1995: 92)

Daphne Spain, in her book *Gendered Spaces* (1992), provides many examples of sex segregation in the home from different cultural contexts and from different time periods. Groups such as the Berbers in Morocco or the Bedouin have a clear spatial division between men's and women's spaces (Figure 8.4).

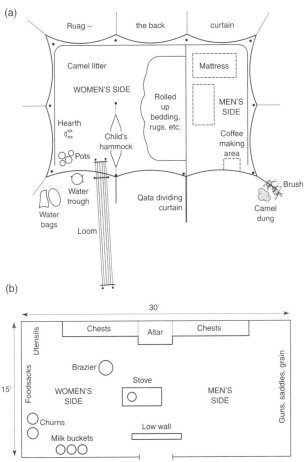

Figure 8.4 Gender divisions of space: (a) Bedouin tent divided by a curtain; (b) Tibetan tent with a symbolic boundary.
Source: Spain, 1992

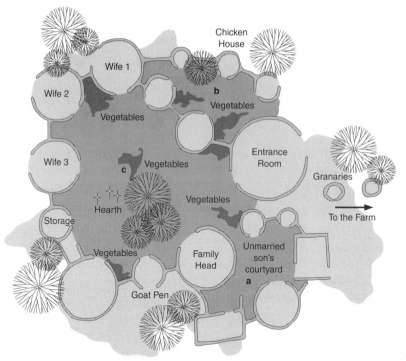

Figure 8.5 Compound household in Nigeria.
Source: Shurma-Smith and Hannam, 1994

Similarly, the Jivaro indians in South America live in tents which have separate entrances for men and women. Shurmer-Smith and Hannam (1994) illustrate a similar point through reference to the Yoruba in Nigeria. Here they point to the non-existence of designated spaces for unmarried women or for children (Figure 8.5). However, such gendered divisions are not confined to the so-called 'Third World'. Within western societies, the idea of the woman's role as home-maker, cook, cleaner, child-rearer, etc. means that she is presumed to 'belong' in some rooms more than others. The distinction in the Victorian home between men in the drawing room discussing serious issues with women confined to the kitchen discussing domestic affairs may seem somewhat antiquated but vestiges of it remain. Again, this reflects a territorial expression of power whereby the designation or apportionment of space within the domestic sphere reflects the status of the individual concerned. Spain reminds us that 'houses are shaped not just by materials and tools, but by ideas, values and norms' (1992: 111). This distinction between the public sphere and the private sphere, mirrored in the architecture of the home, is explained in feminist theory as a mechanism for the subordination of women and the delimiting of their ability to meaningfully enter the public arena. The recognition of territorial behaviour at this level is no less bound up with power than more overt forms of territoriality referred to in previous chapters. As Edward Soja has argued, 'we must be insistently aware of . . . how relations of power and discipline are inscribed into the apparently innocent spatiality of social life' (1989: 6).

It can be argued from a feminist perspective that these divisions of space reflect patriarchy in modes of planning – 'housing and planning policies were shaped towards maintaining an ideal of a family unit in which women were the primary home makers and child rearers' (Roberts 1991: 115). In a sexually stratified labour market women were seen in the role of either home-makers or undertaking particular low-skilled jobs in the paid labour force. This sexual division of labour reflects patriarchal power relations in a 'man-made world'. Power is expressed territorially.

Gay and lesbian spaces

The idea of places territorialized by particular groups on the basis of sexual orientation has been discussed by a number of researchers. At its most elementary level this has seen the mapping of gay and lesbian zones. It is fair to point out that such spaces are not as easy to identify as, say, areas inhabited predominantly by a particular ethnic group. It is equally obvious that, in the main, these are not strictly demarcated areas. Rather, they are zones where gay people may feel more at ease through being accepted rather than rejected, scorned or ignored (or worse) by their neighbours. There have been criticisms of such straightforward mapping of 'gay territories' in that it may be interpreted as a mapping of what some see as deviant behaviour, that it essentializes sexuality and that it reinforces a gay/straight dichotomy (Knopp 1995; Davis 1995). Nevertheless, the fact that gay and lesbian people do, in some instances, cluster in particular places suggests that territorial behaviour may be important. Within these areas, gay-owned businesses, bars and restaurants and a general concentration of gay residents allows the creation of what Castells calls 'a space of freedom' (1997b: 213). Areas such as the Castro district of San Francisco and the more recent evolution of Manchester's 'gay village' serve as important examples of this trend (Plate 8.5).

The construction of such zones may arise for reasons similar to those associated with ghettos and other forms of segregated space. Castells (1997b) has argued that there are two key factors: protection and visibility. The first of these is fairly obvious. Gays or lesbians will feel more comfortable and less vulnerable to attack when surrounded by other gays or lesbians. The idea of 'strength in numbers' may make them feel safer from homophobic 'gaybashers'. The second reason, that of visibility, relates to the need for gay or lesbian people to assert their identity within a culture that is predominantly straight and in which a straight discourse dominates; a society in which homosexuality is still seen by many as deviant or abnormal behaviour. Gay neighbourhoods are a means of allowing homosexuals to assert their identity, a means by which they can say 'we are here'. Harry Britt, a one-time key figure among San Francisco's gay community, once commented that 'when gays are spatially scattered, they are not gay, because they are invisible' (cited in Castells 1997b: 213).

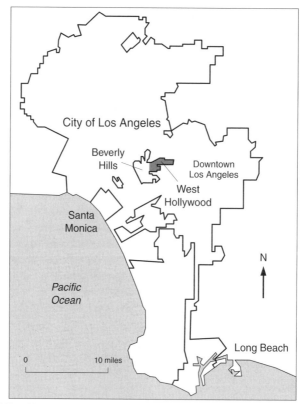

Figure 8.6 West Hollywood, Los Angeles, California.
Source: Forest, 1995

Plate 8.5 Castro District, San Francisco.
Photo courtesy of Jennifer Storey and Fergus Sullivan

Rothenburg (1995) and others have examined the development of lesbian spaces in cities where relatively high concentrations begin to drift into gentrified housing. In this way the place becomes associated with the creation of an identity. As a consequence of the clustering of people sharing that particular identity, the place becomes transformed. In this way a territorial identity emerges which acts as a powerful mechanism in transmitting a sense of the group and in making it easier for gay or lesbian people to feel comfortable with their sexuality and, to some extent at least, encouraging the rest of society to see them in a more positive light.

The significance of San Francisco's gay area was reflected in that community's ability to gain political representation. In obtaining power over territory they also gained political representation and San Francisco has become recognized as something of a 'gay capital' of the United States with a somewhat more liberal attitude exemplified through the hosting of a public annual 'gay wedding day' involving hundreds of couples in a commitment ceremony. In this way the designation of 'gay territories' plays a crucial role in raising awareness of gay people and issues and also provides a means by which they can attain some degree of power and self-confidence. Forest (1995) has demonstrated how designation of a particular place (in this instance West Hollywood in Los Angeles) as a 'gay territory' leads to heightened levels of political involvement and provides some form of legitimation for gay identity, thereby allowing gay people greater control over their own lives (see Box 8.3).

Box 8.3   Gay spaces

The recognition or promotion of certain spaces and places as 'gay areas' can be seen in a variety of ways. Places such as Amsterdam acquire a reputation as havens of tolerance and, indeed, are actively promoted in this way (Binnie 1995). West Hollywood, on the edge of Los Angeles in California, is a small area which in earlier decades had become something of a centre for relatively affluent gay males (Figure 8.6). Its association with gays is reflected in an annual Gay Pride march (inaugurated in 1970) and the founding there of *The Advocate*, now one of the largest gay newsmagazines in the US. By the mid-1980s gays and lesbians were believed to constitute up to 40 per cent of the area's population. It had acquired a reputation as an area tolerant of gays and gay lifestyles with many gay bars and businesses (Forest 1995). Residents endeavoured to have the area incorporated as a separate municipality (see above). Although moves towards municipal incorporation were not totally supported by the gay community initially, a feeling subsequently developed that this presented an opportunity to control their own affairs, having a say in issues that directly affected gays. The campaign was successful and the City of West Hollywood became a reality in 1984 (Moos 1989). Forest (1995) shows there was a clear and concerted attempt to achieve recognition of gays as equal members of civil society. He demonstrates the use of place to obtain a political goal. He argues that the symbolic value of place renders it an effective organizer of identity.

(continued)

(continued)

While this phenomenon of constructing gay neighbourhoods has occurred in places such as West Hollywood, it is important to note that the process may often be intertwined with other forms of neighbourhood or community development. In the Marigny area in New Orleans, the initial gay in-migration was a reflection of middle-class gentrification of an older working-class district. A homogenous 'gay community' did not develop, rather there was a fragmentation along class lines. Knopp (1990) argues that gays were primarily concerned with advancing their class interest through property acquisition rather than a collective gay interest. They found an area in which they were not discriminated against in terms of access to good-quality housing.

At a smaller spatial scale the growing number of gay bars, cafes and shops in places such as Old Compton Street in London's West End made it into a premier gay commercial district. Old Compton Street was temporarily renamed Queer Street during a festival in 1993 (Binnie 1995). While such places may allow a certain degree of freedom for homosexuals, a zone in which they are free to be open about their sexuality, their evolution may be as much to do with commercial interests as sexual freedom. As Binnie (1995) suggests, the place represents an alliance between 'queer politics' and the 'pink pound'. While such spaces may make gay people more self confident, they may also, paradoxically, render those who use them more vulnerable. Known gay bars may become targets for violent attack as shown when a nail bomb in a West End bar in London in early 1999 killed three people.

The examples of gay and lesbian territories suggest another important point, that of the temporality of territory. The longevity of these spaces may be quite brief as the 'scene' moves to somewhere else. Even celebratory events such as 'gay pride' marches (as with women's 'reclaim the night' marches, or as with events such as the Notting Hill carnival in London) can appear as a staking of claim to territory. Thus, these unbounded communities occupy space for a very brief time. Valentine (1995) has drawn attention to the fact that lesbian spaces may be very short lived in time, whether because of the transience of lesbian bars/clubs or the even more short-term phenomena of lesbian or gay evenings. Of course such places and such events may often be 'hijacked' to present an 'acceptable' image of the group concerned. Some activists have expressed disquiet over the appropriation of such events and their dislocation from their original social and cultural roots and from their original territorial base (Hall 1998). Finally, a word of warning. None of this should be used to suggest an essentializing of identities. As Valentine (1995) and others have pointed out there are a multiplicity of lesbian identities just as there are of gay male identities, black identities, national identities, etc.

Home, work and personal space

If territorial behaviour is observable within the urban environment, it can also be identified at an even more micro-scale. Territoriality can be found within the workplace and, indeed, within the home where certain rooms may be seen as the preserve of particular members of the family with others essentially excluded (as suggested earlier in relation to gender roles). In the workplace, divisions between people at different levels in the company hierarchy may be reflected in the allocation of work spaces, offices, floors within buildings and so on.

At its most elementary level, the assertion of territoriality is at its most visible in the claims to private property. Thus, people desire to own their own home, to adorn it in their chosen style and, in various ways, to mark it out as theirs. People, through possession of their own home, can be said to be saying 'this is my territory'. The geographer Jean Gottman suggests that 'civilised people [*sic*] . . . have always partitioned the space around them carefully to set themselves apart from their neighbours' (1973: 1). To some, such as Robert Ardrey, this manifestation is taken as symptomatic of our inherently territorial nature. However, as Alland points out, it might well be the case that 'private property is the child of culture and develops into a major preoccupation only with the evolution of complex society' (1972: 64). It follows that we need to be careful to avoid the trap of translating a need for personal space into an ideological claim for the sanctity of private property.

Even within buildings territorial behaviour can be recognized. The idea of the kitchen as a 'woman's place' is one example of this. As noted earlier, Daphne Spain (1992) has collated details of how the domestic home in many different cultural contexts is often spatially divided in terms of gender and also in terms of age with certain spaces being designated for women or for children and particular spaces seen as male territory. The banning of children from some rooms and the proprietorial attitude towards one's own room in a house are other examples of this. In the home, space is even being claimed at the level of 'my chair', 'my place at the table' and so on.

Equally, in workplaces some areas and rooms can only be entered by staff of a certain level. Some zones are out of bounds to more junior staff. These can be interpreted as managerial strategies designed to ensure a particular outcome; staff know their 'place' and can be more effectively controlled. Hanson and Pratt (1995) reveal how companies reproduce social segregation through spatial practices within the workplace whereby different sets of workers inhabit different parts of the factory and rarely, if ever, meet. Thus, office staff may be located downstairs in 'cubicles' separated by room dividers, with sales staff and management upstairs in individual or shared offices while production staff are located in an entirely separate part of the building. Socializing between workers tends to reflect their spatial segregation even to the extent of each department having separate annual parties. Work hierarchies are reflected in the spatial arrangements of the workplace. These practices have clear outcomes. They will, for example, render it difficult for workers to organize through physically keeping them separate and through engendering a sense of difference between different sections of the workforce.

At a more elementary level, there are echoes of this in how people treat their workspace, whether it be an office, a workstation within an office or whatever. As I type these words I feel quite proprietorial towards not just the equipment I am using, 'my' computer, 'my' books etc., but also towards the office I work in which is seen as 'my' territory, my space in which I do not wish to be disturbed and in which others, while not necessarily being un-welcome, are, nevertheless, only allowed to impinge on my terms. I might let them in but I won't allow them to control 'my' space. In part of course this relates to notions of comfort and familiarity and the desire to protect our own personal space. The psychological need for space and privacy does not, how-ever, take from the fact that what is happening is an attempt to wield power through a territorial mechanism.

Summary

The examples provided in this chapter are evidence of the way in which social relations are expressed through spatial patterns and they highlight the ways in which this geography helps in turn to shape social relations. Phenomena such as racism and sexism regularly embody a territorial component. Territorial strategies are often used to control the 'place' of women or of ethnic or racial groupings. In this way particular ideologies are transposed onto space. People are confronted with wider practices through their use of space or through the ways in which they are allowed to use space. Thus, power relationships take on a spatial dimension, even at the most mundane and ordinary of levels. The basis for territorial division (religion, ethnicity) will vary depending on the particular fault line within that society. The examples used demonstrate the spatialization of wider ideas and they show how people are kept 'in their place' whether through legal means (apartheid), administrative practices (urban incor-poration in the United States) or surveillance strategies (women's fear of crime). Social boundaries are being communicated through space.

However, if dominant systems can be reinforced through territorial behavi-our, they can also be resisted. Territorial strategies are useful mechanisms in the assertion of identity, in making certain groups visible. Spatial concentrations within particular territories, sometimes loosely defined, make visible people and issues which might otherwise remain unseen. In doing so, this demonstrates the 'positive' and 'negative' dimensions to territoriality; it can be both a force for oppression and also one for liberation. Particular strategies can be used to assert an identity and territorially transgressive acts can be employed to reclaim space. So-called 'reclaim the night' marches by women are an act of resistance to the forces which make women fearful of certain public spaces. In a broader sense the storming of certain bastions of male–white power can be seen as attempts to wrest space and, hence, power from the hands of those currently in control.

Further reading

On residential segregation, see the relevant sections of urban geography texts such as H. Carter, *The Study of Urban Geography* (4th edition, Arnold, London, 1995), P. Knox, *Urban Social Geography. An Introduction* (3rd edition, Longman, Harlow, 1995), J.R. Short, *The Urban Order. An Introduction to Cities, Culture and Power* (Blackwell,

Cambridge MA, 1996) and D.T. Herbert and C.J. Thomas, *Cities in Space. City as Place* (David Fulton, London, 1990). On specific issues of ethnic segregation see F.W. Boal, 'Ethnic residential segregation', in D.T. Herbert and R.J. Johnston (eds), *Social Areas in Cities* (Wiley, London, 1978, pp. 57–95), Peter Jackson (ed.), *Race and Racism. Essays in Social Geography* (Allen and Unwin, London, 1987) and the same author's *Maps of Meaning. An Introduction to Cultural Geography* (Routledge, London, 1989). On the specifics of ghetto formation, see various articles by Ceri Peach including 'Does Britain have ghettos?', *Transactions of the Institute of British Geographers* NS 1996, **21**(1): 216–35.

For a good overview of apartheid from a geographical perspective see the work of D.M. Smith, most notably his edited volume *The Apartheid City and Beyond. Urbanization and Social Change in South Africa* (Routledge, London, 1992). See also J. Dixon, S. Reicher and D.H. Foster, 'Ideology, geography, racial exclusion', *Text* 1997, **17**(3): 317–48. The standard work on territoriality in Belfast is that by F.W. Boal which appears in a number of publications including 'Territoriality on the Shankill–Falls divide, Belfast', *Irish Geography* 1969, **6**(1): 30–50. See also various chapters in *Integration and Division. Geographical Perspectives on the Northern Ireland Problem*, edited by F.W. Boal and J.N.H. Douglas (Academic Press, London, 1982), Emrys Jones, *A Social Geography of Belfast* (Oxford University Press, London, 1960) and P. Doherty, 'Ethnic segregation levels in the Belfast urban area', *Area* 1989, **21**(2): 151–9. On murals and other symbols, see the various articles in A.D. Buckley (ed.), *Symbols in Northern Ireland* (Institute of Irish Studies, Queens University, Belfast, 1998). The standard overview of the processes of residential segregation in the United States is R.J. Johnston, *Residential Segregation, the State and Constitutional Conflict in American Urban Areas* (Academic Press, London, 1984). On Los Angeles, see Mike Davis, *City of Quartz. Excavating the Future in Los Angeles* (Verso, London, 1990). One of the few works to deal with graffiti in its geographical–territorial context is David Ley and Roman Cybriwsky 'Urban graffiti as territorial markers', *Annals of the Association of American Geographers* 1974, **64**(4): 491–505.

For discussions of geography and gender see Doreen Massey, *Space, Place and Gender* (Polity Press, Cambridge, 1994). For a good collation of material on gendered domestic spaces see Daphne Spain, *Gendered Spaces* (University of North Carolina Press, Chapel Hill NC, 1992). More broadly see Gill Valentine, 'The geography of women's fear', *Area* 1989, **21**(4): 385–90. See also the relevant chapters in P. Shurmer-Smith and K. Hannam, *Worlds of Desire, Realms of Power. A Cultural Geography* (Edward Arnold, London, 1994).

For a range of perspectives on gay and lesbian spaces see various chapters in David Bell and Gill Valentine (eds), *Mapping Desire. Geographies of Sexualities* (Routledge, London, 1995). On West Hollywood, see B. Forest, 'West Hollywood as symbol: the significance of place in the construction of a gay identity', *Environment and Planning D: Society and Space* 1995, **13** (2): 133–57.

Chapter 9

Conclusions

This book has presented a range of examples of human territorial formations and territorial strategies drawn from a variety of spatial scales ranging from the global to the local. Central to the examples used has been a highlighting of:

1. the way in which particular ideologies and social practices are manifested in space,
2. the importance of territory as a component of self-identity and, more significantly, group identity,
3. the employment of territorial strategies to gain or to maintain power,
4. the utilization of territorial strategies or territorial frames in order to actively resist the imposition of power by dominant groups.

Much attention was devoted to the most obvious form of territorial division, the state system. In considering the evolution and sustaining of this system there was a particular focus on the territorial ideology of nationalism, a key building block in the formation of present-day states. While these territorial divisions, and numerous formalized substate territorial divisions, can be seen as spatial containers, they are much more than this. The existence of discrete territorial units, and the boundaries between them, reflects the world of politics where control over territory can be seen to signify power. The division of geographic space into territories, at whatever scale, represents the spatial expression of power. Despite various forces working to alter or to reduce the extent of state sovereignty, often subsumed under the umbrella of globalization, issues of control over territory are likely to continue to be of significance.

While the system of territorial states is the most obvious form of territoriality, numerous more micro-scale examples also occur. Apart from administrative subdivisions and various forms of community-based territorializations, there are also many less formalized attempts to demarcate geographic space. Some people may be excluded, or at least attempts may be made to exclude them, from certain spaces on the basis of 'race', ethnicity, gender, sexual orientation or simply because they belong to a category seen as 'other'. These various manifestations of territoriality can be seen as power expressed through a territorial frame.

Throughout the book a key idea has been that territorial strategies are useful organizing devices and, more significantly, they are a means through which power is maintained or contested. From the state downwards, territories are constructed and used as means of social, economic and political control. It follows from this that territories cannot be seen as naturally occurring entities; rather they are human creations. Thus, territorial behaviour in humans is not an innate tendency, rather it is a product of social, cultural, economic and political circumstances. While recognizing the constructed nature of territoriality and the functions it serves, the importance of territorial arrangements in peoples' everyday lives cannot be ignored. It is obvious that people form attachments to place and these can have a huge impact on how people think and act. The significance of nationalism, an ideology based on attachment to a 'homeland', is a prime example. Place undoubtedly matters.

Territorial control is not a passive phenomenon. Particular territorial constructions are resisted, as in secessionist nationalism, opposition to the EU or disputes surrounding squatter settlements in South Africa. From the Zapatistas in Chiapas to 'reclaim the night' marches, control of territory is subject to constant struggle as attempts are made to control space or to wrest space from other groups. All forms of human territorial behaviour reflect power relations. Territorial control and the contestations over geographic space reflect the workings of particular political processes. It follows that geographers and others must explore the underlying causes, not merely map the territorial manifestations. As Sack (1986) observes, an emphasis on territoriality can tend to obscure the real actors. By focusing on the units (whether formalized and rigidly demarcated or whether informal and hazily bounded) attention is switched away from the protagonists and onto the end product of their activity. It is the processes giving rise to the patterns that are of ultimate importance in determining the welfare of those living within particular defined spaces. It is the mechanisms underpinning the territorializations, rather than the territories themselves, that require investigation.

Bibliography

Agnew, J., 1994, The territorial trap: the geographical assumptions of international relations theory, *Review of International Political Economy* 1(1): 53–80.

Agnew, J., 1995, Postscript: federalism in post-Cold War era, in G. Smith (ed.) *Federalism. The Multiethnic Challenge*, London: Longman, pp. 294–302.

Agnew, J. (ed.), 1997, *Political Geography. A Reader*, London: Arnold.

Agnew, J., 1998, *Geopolitics. Revisioning World Politics*, London: Routledge.

Agnew, J. and S. Corbridge, 1995, *Mastering Space. Hegemony, Territory and International Political Economy*, London: Routledge.

Aitchison, J. and H. Carter, 1994, *A Geography of the Welsh Language, 1961–1991*, Cardiff: University of Wales Press.

Alland, A. Jr, 1972, *The Human Imperative*, New York: Columbia University Press.

Allen, J. and D. Massey (eds), 1995, *Geographical Worlds*, Oxford: Oxford University Press/Open University.

Alter, P., 1989, *Nationalism*, London: Edward Arnold.

Anderson, B., 1991, *Imagined Communities. Reflections on the Origin and Spread of Nationalism*, revised edition, London: Verso.

Anderson, J., 1995, The exaggerated death of the nation-state, in J. Anderson, C. Brook and A. Cochrane (eds) *A Global World? Re-ordering Political Space*, Oxford: Open University/Oxford University Press, pp. 65–112.

Anderson, J., 1996, The shifting stage of politics: new medieval and postmodern territorialities?, *Environment and Planning D: Society and Space* 14(2): 133–53.

Anderson, J., C. Brook and A. Cochrane (eds), 1995, *A Global World? Re-ordering Political Space*, Oxford: Open University/Oxford University Press.

Anderson, M., 1996, *Frontiers. Territory and State Formation in the Modern World*, Cambridge: Polity Press.

Ardrey, R., 1967, *The Territorial Imperative. A Personal Inquiry into the Animal Origins of Property and Nations*, London: Collins.

Ascherson, N., 1996, *Black Sea. The Birthplace of Civilisation and Barbarism*, London: Vintage.

Bachrach, P. and M. Baratz, 1962, Two faces of power, *American Political Science Review* 56: 947–52.

Balakrishnan, G. (ed.), 1996, *Mapping the Nation*, London: Verso.

Banfield, E.C., 1958, *The Moral Basis of a Backward Society*, New York: Free Press.

Barbalet, J.M., 1988, *Citizenship. Rights, Struggle and Class Inequality*, Milton Keynes: Open University Press.

Baubock, R., 1994, *Transnational Citizenship. Membership and Rights in International Migration*, Aldershot: Edward Elgar.

Bell, D. and G. Valentine (eds), 1995, *Mapping Desire. Geographies of Sexualities*, London: Routledge.

Billig, M., 1995, *Banal Nationalism*, London: Sage.

Binnie, J., 1995, Trading places: consumption, sexuality and the production of queer space, in D. Bell and G. Valentine (eds) *Mapping Desire. Geographies of Sexualities*, London: Routledge, pp. 182–99.

Blaut, J.M., 1987, *The National Question*, London: Zed Books.

Boal, F.W., 1969, Territoriality on the Shankill–Falls divide, Belfast, *Irish Geography* 6(1): 30–50.

Boal, F.W., 1978, Ethnic residential segregation, in D.T. Herbert and R.J. Johnston (eds) *Social Areas in Cities*, London: Wiley, pp. 57–95.

Boal, F.W. and J.N.H. Douglas (eds), 1982, *Integration and Division. Geographical Perspectives on the Northern Ireland Problem*, London: Academic Press.

Bogdanor, V., 1999, Devolution: decentralisation or disintegration?, *The Political Quarterly* 70(2): 185–94.

Boyce, D.G., 1982, *Nationalism in Ireland*, London: Croom Helm.

Bozzoli, B., 1999, Space and identity in rebellion. Power, target, resource. Paper presented at Regional Conference on Social Movements and Change, Cork.

Brook, C., 1995, The drive to global regions, in J. Anderson, C. Brook and A. Cochrane (eds) *A Global World? Re-ordering political space*, Oxford: Open University Press, pp. 113–65.

Buckley, A.D. (ed.), 1998, *Symbols in Northern Ireland*, Belfast: Institute of Irish Studies, Queens University.

Bunce, M., 1994, *The Countryside Ideal. Anglo-American Images of Landscape*, London: Routledge.

Byrne, L., 1996, *Local Government Transformed*, Manchester: Baseline Books.

Carson, C., 1998, *The Star Factory*, London, Granta.

Carter, H., 1995, *The Study of Urban Geography*, 4th edition, London: Arnold.

Castells, M., 1997a, *The Rise of the Network Society*, Malden MA: Blackwell.

Castells, M., 1997b, *The Power of Identity*, Malden MA: Blackwell.

Castells, M., 1997c, *End of Millennium*, Malden MA: Blackwell.

Cesarani, D. and M. Fulbrook (eds), 1996, *Citizenship, Nationality and Migration in Europe*, London: Routledge.

Chatterjee, P., 1993, *The Nation and its Fragments. Colonial and Post-Colonial Histories*, Cambridge: Cambridge University Press.

Chatterjee, P., 1995, Whose imagined community?, in G. Balakrishnan (ed.), *Mapping the Nation*, London: Verso, pp. 214–25.

Clark, G.L. and M. Dear, 1984, *State Apparatus. Structures of Language and Legitimacy*, Boston MA: Allen & Unwin.

Cloke, P., C. Philo and D. Sadler, 1991, *Approaching Human Geography. An Introduction to Contemporary Theoretical Debates*, London: Paul Chapman.

Cloke, P., P. Crang and M. Goodwin (eds), 1999, *Introducing Human Geographies*, London: Arnold.

Close, P., 1995, *Citizenship, Europe and Change*, Basingstoke: Macmillan.

Cochrane, A., 1993, *Whatever Happened to Local Government?*, Buckingham: Open University Press.

Collins, R., 1990, *The Basques*, 2nd edition, Oxford: Blackwell.

Connor, W., 1978, A nation is a nation, is a state is an ethnic group is a . . . , *Ethnic and Racial Studies* 1(4): 377–400.

Cosgrove, D. and S. Daniels (eds), 1998, *The Iconography of Landscape*, Cambridge: Cambridge University Press.

Cox, K., 1991, Comment. Redefining 'territory', *Political Geography Quarterly* **10**(1): 5–7.

Cullingworth, J.B. and V. Nadin, 1997, *Town and Country Planning in the UK*, 12th edition, London: Routledge.

Curry, N., 1993, Rural development in the 1990s – does prospect lie in retrospect?, in M. Murray and J. Greer (eds) *Rural Development in Ireland: a Challenge for the 1990s*, Aldershot: Avebury, pp. 21–39.

Daniels, S., 1993, *Fields of Vision. Landscape Imagery and National Identity in England and the United States*, Cambridge: Polity Press.

da Silva, K.M., 1996, Ethnicity and nationalism, in L. van de Goor, K. Rupesinghe and P. Sciarone (eds) *Between Development and Destruction. An Enquiry into the Causes of Conflict in Post-Colonial States*, The Hague: Netherlands Ministry of Foreign Affairs, pp. 109–25.

Davidson, B., 1992, *The Black Man's Burden. Africa and The Curse of the Nation-State*, Oxford: James Currey.

Davis, M., 1990, *City of Quartz: Excavating the Future in Los Angeles*, London: Verso.

Davis, T., 1995, The diversity of queer politics and the redefinition of sexual identity and community in urban spaces, in D. Bell and G. Valentine (eds) *Mapping Desire. Geographies of Sexualities*, London: Routledge, pp. 284–303.

Dawkins, R., 1976, *The Selfish Gene*, Oxford: Oxford University Press.

Dent, M., 1995, Ethnicity and Territorial Politics in Nigeria, in C. Smith (ed.) *Federism: the Multiethnic Challenge*, London: Longman, pp. 128–53.

DeSipio, L. and R.O. de la Garza 1998, *Making Americans, Remaking America: Immigration and Immigrant Policy*, Boulder CO: Westview Press.

Dicken, P., 1998, *Global Shift. Transforming the World Economy*, 3rd edition, London: Paul Chapman.

Dijkink, G., 1996, *National Identity and Geopolitical Visions. Maps of Pride and Pain*, London: Routledge.

Dixon, J. A. and K. Durrheim, 2000, Displacing place identity: a discursive approach to locating self and other, *British Journal of Social Psychology* **39**(1): 27–44.

Dixon, J.A., S. Reicher and D.H. Foster, 1997, Ideology, geography, racial exclusion, *Text* **17**(3): 317–48.

Dobson, A., 1995, *Green Political Thought*, 2nd edition, London: Unwin Hyman.

Dodds, K., 1998, Political geography I: the globalization of world politics, *Progress in Human Geography* **22**(4): 595–606.

Dodds, K., 2000, *Geopolitics in a Changing World*, Harlow: Prentice Hall.

Doherty, P., 1989, Ethnic segregation levels in the Belfast urban area, *Area* **21**(2): 151–9.

Dorling, D. and D. Fairbairn, 1997, *Mapping. Ways of Representing the World*, Harlow: Addison Wesley Longman.

Doyle, T. and D. McEachern, 1998, *Environment and Politics*, London: Routledge.

Duignan, P. and L. Gann, 1985, *The United States and Africa: A History*, Cambridge: Cambridge University Press.

Dwyer, C., 1999, Migrations and diasporas, in P. Cloke, P. Crang and M. Goodwin (eds) *Introducing Human Geographies*, London: Arnold, pp. 287–95.

Ericksen, E.G., 1980, *The Territorial Experience. Human Ecology as Symbolic Interaction*, Austin TX: University of Texas Press.

Ethics, Place and Environment 2(2), 1999.

Evans, R., 1995, Brave new world order, *Geographical* Vol. LXVII, No. 1.

Eyles, J., 1985, *A Sense of Place*, Warrington: Silverbrook Press.

Faegre, T., 1979, *Tents: Architecture of the Nomads*, New York: Doubleday.

Featherstone, M. (ed.), 1990, *Global Culture: Nationalism, Globalisation and Modernity*, London: Sage.

Fell, A., 1991, Penthesilea, perhaps, in M. Fisher and U. Owen (eds) *Whose Cities?*, London: Penguin, pp. 73–84.

Fieldhouse, D.K., 1982, *The Colonial Empires. A Comparative Survey from the Eighteenth Century*, Basingstoke: Macmillan.

Fonseca, I., 1996, *Bury Me Standing. The Gypsies and their Journey*, London: Vintage.

Forest, B., 1995, West Hollywood as symbol: the significance of place in the construction of a gay identity, *Environment and Planning D: Society and Space* 13(2): 133–57 (excerpts reproduced in L. McDowell, (ed.), 1997, *Undoing Place: A Geographical Reader*, London: Arnold, pp. 112–29).

Forrest, J.B., 1988, The quest for state hardness in Africa, *Comparative Politics* 20: 423–41.

Foucault, M., 1977, *Discipline and Punish. The Birth of the Prison*, London: Penguin.

Foucault, M., 1980, *Power/Knowledge. Selected Interviews and Other Writings* (ed. C. Gordon), Brighton: Harvester Press.

Fukuyama, F., 1992, *The End of History and The Last Man*, New York: Free Press.

Gamberale, C., 1997, European citizenship and political identity, *Space and Polity* 1(1): 37–59.

Gellner, E., 1983, *Nations and Nationalism*, Oxford: Blackwell.

Gellner, E., 1994, *Encounters with Nationalism*, Oxford: Blackwell.

Gellner, E., 1997, *Nationalism*, London: Weidenfeld and Nicolson.

Giddens, A., 1990, *The Consequences of Modernity*, Cambridge: Polity Press.

Gilbert, D., 1999, Sponsorship, academic independence and critical engagement: a forum on Shell, the Ogoni dispute and the Royal Geographical Society (with the Institute of British Geographers). Introduction, *Ethics, Place and Environment* 2(2): 219–28.

Glassner, M.I., 1993, *Political Geography*, New York: Wiley.

Glenny, M., 1996, *The Fall of Yugoslavia. The Third Balkan War*, 3rd edition, Harmondsworth: Penguin.

Gold, J.R., 1982, Territoriality and human spatial behaviour, *Progress in Human Geography* 6(1): 44–67.

Goldblatt, D., D. Held, A. McGrew and J. Perraton, 1997, Economic globalisation and the nation-state: shifting balances of power, *Alternatives* 22: 269–85.

Goldenberg, S., 1994, *Pride of Small Nations. The Caucasus and Post-Soviet Disorder*, London: Zed Books.

Gottman, J., 1973, *The Significance of Territory*, Charlottesville VA: University Press of Virginia.

Gould, S.J., 1983, *The Mismeasure of Man*, London: Penguin.

Gould, S.J., 1991, *The Flamingo's Smile*, London: Penguin.

Graham, B., 1994, No place of the mind: contested Protestant representations of Ulster, *Ecumene* 1(3): 257–81.

Graham, B. (ed.), 1997, *In Search of Ireland. A Cultural Geography*, London: Routledge.

Graham, B. (ed.), 1998, *Modern Europe. Place, Culture and Identity*, London: Arnold.

Gramsci, A., 1971, *Selections from the Prison Notebooks*, New York: International Publishers.

Grant, L., 1998, In the ghetto, *Guardian* (Weekend), 25 July, pp. 16–22.

Gray, C., 1994, *Government Beyond the Centre. Sub-National Politics in Britain*, London: Macmillan.

Gruffudd, P., 1995, Remaking Wales: nation-building and the geographical imagination, 1925–50, *Political Geography* 14(3): 219–39.

Guibernau, M., 1995, Spain: a federation in the making?, in G. Smith (ed.) *Federalism. The Multiethnic Challenge*, London: Longman, pp. 239–54.

Guibernau, M., 1996, *Nationalism. The Nation-State and Nationalism in the Twentieth Century*, Cambridge: Polity Press.

Hall, E.T., 1959, *The Silent Language*, Garden City: Doubleday.

Hall, T., 1998, *Urban Geography*, London: Routledge.

Hanson, S. and G. Pratt, 1995, *Gender, Work and Space*, London: Routledge.

Harley, J.B., 1988, Maps, knowledge and power, in D. Cosgrove and S. Daniels (eds) *The Iconography of Landscape*, Cambridge: Cambridge University Press, pp. 277–312 (excerpts reproduced in Daniels, S. and R. Lee (eds), 1996, *Exploring Human Geography. A Reader*, London: Arnold, pp. 377–94).

Hartshorne, R., 1950, The functional approach in political geography, *Annals of the Association of American Geographers* 40(1): 95–130.

Harvey, D., 1989, *The Condition of Postmodernity*, Oxford: Blackwell.

Harvey, D., 1996, *Justice, Nature and the Geography of Difference*, Oxford: Blackwell.

Hay, C., 1996, *Re-stating Social and Political Change*, Buckingham: Open University Press.

Hechter, M., 1975, *Internal Colonialism: The Celtic Fringe in British National Development*, London: Routledge.

Held, D., 1989, *Political Theory and the Modern State*, Cambridge: Polity Press.

Held, D., J. Anderson, B. Gieben, S. Hall, L. Harris, P. Lewis, N. Parker and B. Turok (eds), 1985, *States and Societies*, Oxford: Basil Blackwell.

Herbert, D.T. and C.J. Thomas, 1990, *Cities in Space. City as Place*, London: David Fulton.

Hill, D.M., 1994, *Citizens and Cities. Urban Policy in the 1990s*, Hemel Hempstead: Harvester Wheatsheaf.

Hinde, R.A., 1987, *Individuals, Relationships and Culture: Links Between Ethology and the Social Sciences*, Cambridge: Cambridge University Press.

Hobsbawm, E., 1992, *Nations and Nationalism since 1780. Programme, Myth, Reality*, 2nd edition, Cambridge: Cambridge University Press.

Hobsbawm, E., 1998, *On History*, London: Abacus.

Hobsbawm, E. and T. Ranger (eds), 1992, *The Invention of Tradition*, Cambridge: Cambridge University Press.

Hoffman, J., 1995, *Beyond the State. An Introductory Critique*, Cambridge: Polity Press.

Ingold, T., 1986, *The Appropriation of Nature. Essays on Human Ecology and Social Relations*, Manchester: Manchester University Press.

Jackson, P. (ed.), 1987, *Race and Racism. Essays in Social Geography*, London: Allen and Unwin.

Jackson, P., 1989, *Maps of Meaning. An Introduction to Cultural Geography*, London: Routledge.

James, P.E., 1959, *Latin America*, 3rd edition, New York: Cassell.

Johnson, N., 1995, Cast in stone: monuments, geography and nationalism, *Environment and Planning D: Society and Space* 13(1): 51–65 (reproduced in J. Agnew (ed.), 1997, *Political Geography. A Reader*, London: Arnold, pp. 347–64).

Johnston, R.J., 1982, *The American Urban System: a Geographical Perspective*, London: Longman.

Johnston, R.J., 1984, *Residential Segregation, the State and Constitutional Conflict in American Urban Areas*, London: Academic Press.

Johnston, R.J., 1991a, *A Question of Place. Explaining the Practice of Geography*, Oxford: Blackwell.

Johnston, R.J., 1991b, *Geography and Geographers. Anglo-American Human Geography since 1945*, 4th edition, London: Edward Arnold.

Jones, E., 1960, *A Social Geography of Belfast*, London: Oxford University Press.

Journal of Rural Studies 14(1), 1998.

Kellas, J.G., 1991, *The Politics of Nationalism and Ethnicity*, Basingstoke: Macmillan.

Khan, S., 1994, *Nigeria. The Political Economy of Oil*, Oxford: Oxford University Press.

Kinsman, P., 1995, Landscape, race and national identity: the photography of Ingrid Pollard, *Area* 27(4): 300–10.

Kirby, P., 1988, *Has Ireland a Future?*, Cork: Mercier Press.

Kitchin, R. and N.J. Tait, 2000, *Conducting Research in Human Geography: Theory, Methodology and Practice*, Harlow: Prentice Hall.

Kliot, N. and Y. Mansfield, 1997, The Political Landscape of Partition: The Case of Cyprus, *Political Geography* 16(6): 495–521.

Knopp, L., 1990, Some theoretical implications of gay involvement in an urban land market, *Political Geography Quarterly* 9(4): 337–52.

Knopp, L., 1995, Sexuality and urban space: a framework for analysis, in D. Bell and G. Valentine (eds) *Mapping Desire. Geographies of Sexualities*, London: Routledge, pp. 149–61.

Knox, P., 1995, *Urban Social Geography. An Introduction*, 3rd edition, Harlow: Longman.

Knox, P. and J. Agnew, 1994, *The Geography of the World Economy*, 2nd edition, London: Edward Arnold.

Kohn, H., 1967, *The Idea of Nationalism*, 2nd edition, New York: Collier-Macmillan.

Kuper, S., 1994, *Football Against the Enemy*, London: Orion.

Laitin, D.D., C. Solé and S.N. Kalyvas, 1994, Language and the construction of states: the case of Catalonia in Spain, *Politics and Society* 22(1): 5–29.

Lewis, O., 1966, The culture of poverty, *Scientific American* 215(4): 19–25.

Ley, D. and R. Cybriwsky, 1974, Urban graffiti as territorial markers, *Annals of the Association of American Geographers* 64(4): 491–505.

Livingstone, D.N., 1992, *The Geographical Tradition. Episodes in the History of a Contested Enterprise*, Oxford: Blackwell.

Lorenz, K., 1966, *On Aggression*, New York: Harcourt, Brace and World.

Lowenthal, D., 1994, European and English landscapes as national symbols, in D. Hooson (ed.) *Geography and National Identity*, Oxford: Blackwell, pp. 15–38.

Lowenthal, D., 1998, *The Heritage Crusade and the Spoils of History*, Cambridge: Cambridge University Press.

MacLaughlin, J., 1986, The political geography of 'nation-building' and nationalism in social sciences: structural vs. dialectical accounts, *Political Geography Quarterly* 5(4): 299–329.

Major, J., 1991, quoted in *Observer*, 30 June.

Malmberg, T., 1980, *Human Territoriality. Survey of Behavioural Territories in Man with Preliminary Analysis and Discussion of Meaning*, The Hague: Mouton.

Manent, P., 1997, Democracy without nations?, *Journal of Democracy* 8(2): 92–102.

Mann, M., 1984, The autonomous power of the state, *European Journal of Sociology* **25**(2): 185–213 (reproduced in Agnew, J. (ed.), 1997, *Political Geography. A Reader*, London: Arnold, pp. 58–81).

Manzoni, M. and P. Pagnini, 1996, Comment. The symbolic territory of Antarctica, *Political Geography* **15**(5): 359–64.

Marqusee, M., 1995, Fear and fervour, *Guardian* (G2), 4 July, p. 2.

Marshall, T.H., 1950, *Citizenship and Social Class*, Cambridge: Cambridge University Press.

Marx, K. and F. Engels, 1969, *The Manifesto of the Communist Party*, Moscow: Progress Publishers.

Massey, D., 1994, *Space, Place and Gender*, Cambridge: Polity Press.

May, J., 1996, Globalization and the politics of place: place and identity in an inner London neighbourhood, *Transactions of the Institute of British Geographers, New Series* **21**(1): 194–215.

McAuley, J.W., 1994, *The Politics of Identity. A Loyalist Community in Belfast*, Aldershot: Avebury.

McDowell, L., 1983, Towards an understanding of the gender division of urban space, *Environment and Planning D: Society and Space* **1**(1): 59–72.

McDowell, L. and D. Massey, 1984, A woman's place, in D. Massey and J. Allen (eds) *Geography Matters!, A Reader*, Cambridge: Cambridge University Press, pp. 128–47.

Meegan, R., 1995, Local Worlds, in J. Allen and D. Massey (eds), *Geographical Worlds*, Oxford: Open University/Oxford University Press, pp. 53–104.

Meek, J., 1999, One people cut apart by Ikea curtain, *Guardian*, 4 February, p. 14.

Miliband, R., 1969, *The State in Capitalist Society*, London: Quartet.

Miliband, R., 1991, Reflections on the crisis of communist regimes, in R. Blackburn (ed.), *After the Fall. The Failure of Communism and the Future of Socialism*, London: Verso, pp. 6–17.

Miller, D., 1995, *On Nationality*, Oxford: Clarendon Press.

Moos, A., 1989, The grassroots in action: gays and seniors capture the local state in West Hollywood, California, in J. Wolch and M. Dear (eds), *The Power of Geography. How Territory Shapes Social Life*, Boston MA: Unwin Hyman, pp. 351–69.

Morris, D., 1973, *Manwatching. A Field Guide to Human Behaviour*, London: Jonathan Cape.

Morris, D., 1981, *The Soccer Tribe*, London: Jonathan Cape.

Morris, D., 1994, *The Naked Ape. A Zoologist's Study of the Human Animal*, London: Vintage.

Mosca, G., 1939, *The Ruling Class*, New York: McGraw-Hill.

Muir, R., 1997, *Political Geography. A New Introduction*, Basingstoke: Macmillan.

Murdoch, J., 1997, The shifting territory of government, *Area* **29**(2): 109–18.

Murphy, A., 1993, Linguistic regionalism and the social construction of space in Belgium, *International Journal of the Sociology of Language* **104**(1): 49–64 (reproduced in Agnew, J. (ed.), 1997, *Political Geography. A Reader*, London: Arnold, pp. 256–69).

Murphy, A., 1995, Belgium's regional divergence: along the road to federation, in G. Smith (ed.) *Federalism. The Multiethnic Challenge*, London: Longman, pp. 73–100.

Nairn, T., 1977, *The Break-up of Britain*, London: New Left Books.

Nairn, T., 1988, *The Enchanted Glass. Britain and its Monarchy*, London: Chandos.

Nairn, T., 1997, *Faces of Nationalism. Janus Revisited*, London: Verso.

Nash, C., 1993, 'Embodying the nation': the west of Ireland landscape and Irish identity, in B. O'Connor and M. Cronin (eds) *Tourism in Ireland. A Critical Analysis*, Cork: Cork University Press, pp. 86–112.

Newman, D. and A. Paasi, 1998, Fences and neighbours in the postmodern world: boundary narratives in political geography, *Progress in Human Geography* 22(2): 186–207.

Nodia, G., 1994, Nationalism and democracy, in L. Diamond and M.F. Plattner (eds) *Nationalism, Ethnic Conflict and Democracy*, Baltimore MD: Johns Hopkins University Press, pp. 3–22.

O'Dowd, L., 1996, British nationalism and the impasse over the peace process. Paper presented at the Annual Conference of the British Sociological Association, Reading.

Osei-Kwame, P. and P.J. Taylor, 1984, A politics of failure: the political geography of Ghanaian elections, 1954–1979, *Annals of the Association of American Geographers* 74(4): 574–89 (reproduced in Agnew, J. (ed.), 1997, *Political Geography. A Reader*, London: Arnold, pp. 198–219).

Ó Tuathail, G., 1996, *Critical Geopolitics. The Politics of Writing Global Space*, London, Routledge.

Ó Tuathail, G., 1998, Political geography III: dealing with deterritorialization, *Political Geography* 22(1): 81–93.

Ó Tuathail, G., S. Dalby and P. Routledge (eds), 1998, *The Geopolitics Reader*, London: Routledge.

Paasi, A., 1991, *Territories, Boundaries and Consciousness: The Changing Geographies of the Finnish–Russian Border*, Chichester: Wiley.

Paasi, A., 1999, Boundaries as social practice and discourse: the Finnish–Russian border, *Regional Studies* 33(7): 669–80.

Peach, C., 1996, Does Britain have ghettos?, *Transactions of the Institute of British Geographers, New Series* 21(1): 216–35.

Peet, R., 1998, *Modern Geographical Thought*, Oxford: Blackwell.

Pepper, D., 1996, *Modern Environmentalism: An Introduction*, London: Routledge.

Philo, C., 1989, 'Enough to drive one mad': the organization of space in 19th-century lunatic asylums, in J. Wolch and M. Dear (eds) *The Power of Geography. How Territory Shapes Social Life*, Boston MA: Unwin Hyman, pp. 258–90.

Piaget, J. and B. Inhelder, 1967, *The Child's Conception of Space*, New York: Norton.

Pile, S. and M. Keith (eds), 1997, *Geographies of Resistance*, London: Routledge.

Plamenatz, J., 1976, Two types of nationalism, in E. Kamenka (ed.) *Nationalism: The Nature and Evolution of an Idea*, London: Edward Arnold, pp. 22–36.

Political Geography 14(2), 1995 (special issue: Spaces of Citizenship).

Political Geography 17(2), 1998 (special issue: Space, Place and Politics in Northern Ireland).

Political Studies 42: 1994 (special issue: Contemporary Crisis of the Nation-State).

Popovski, V., 1995, Yugoslavia: politics, federation, nation, in G. Smith (ed.) *Federalism. The Multiethnic Challenge*, London: Longman, pp. 180–207.

Poulantzas, N., 1969, The problem of the capitalist state, *New Left Review* 58: 119–33.

Prescott, J.R.V., 1987, *Political Frontiers and Boundaries*, London: Unwin Hyman.

Radcliffe, S.A., 1998, Frontiers and popular nationhood: geographies of identity in the 1995 Ecuador–Peru border dispute, *Political Geography* 17(3): 273–93.

Ray, C., 1998, Territory, structures and interpretation – two case studies of the European Union's LEADER I programme, *Journal of Rural Studies* 14(1): 79–87.

Regional Studies 33(7), 1999 (special edition: State Borders and Border Regions).

Roche, M., 1992, *Rethinking Citizenship. Welfare, Ideology and Change in Modern Society*, Cambridge: Polity Press.

Roberts, M., 1991, *Living in a Man-Made World. Gender Assumptions in Modern Housing Design*, London: Routledge.

Rolston, B., 1997–98, From King Billy to Cú Chulainn: loyalist and republican murals, past, present and future, *Éire–Ireland* (Winter/Spring/Summer): 6–28.

Rose, G., 1995, Place and identity: a sense of place, in D. Massey and P. Jess (eds) *A Place in the World? Places, Culture and Globalization*, Oxford: Oxford University Press/Open University, pp. 87–132.

Rose, S., L.J. Kamin and R.C. Lewontin 1990, *Not in our Genes: Biology, Ideology and Human Nature*, Harmondsworth: Penguin.

Rothenberg, T., 1995, 'And she told two friends': lesbians creating urban social space, in D. Bell and G. Valentine (eds) *Mapping Desire. Geographies of Sexualities*, London: Routledge, pp. 165–81.

Rousseau, J.J., 1973, *The Social Contract and Discourses*, London: Dent.

Ruane, J. and J. Todd, 1996, *The Dynamics of Conflict in Northern Ireland. Power, Conflict and Emancipation*, Cambridge: Cambridge University Press.

Rumley, D. and J. Minghi (eds), 1991, *The Geography of Border Landscapes*, London: Routledge.

Rushdie, S., 1992, *Imaginary Homelands. Essays and Criticism 1981–1991*, London, Granta.

Rutter, M. and N. Madge, 1976, *Cycles of Disadvantage*, London: Heinemann.

Sack, R., 1983, Human territoriality: a theory, *Annals of the Association of American Geographers* 73(1): 55–74.

Sack, R., 1986, *Human Territoriality: Its Theory and History*, Cambridge: Cambridge University Press.

Said, E.W., 1992, *The Question of Palestine*, new edition, London: Vintage.

Said, E.W., 1995, *Orientalism. Western Conceptions of the Orient*, London: Penguin.

Schumpeter, J.A., 1976, *Capitalism, Socialism and Democracy* (5th edition), London: George Allen and Unwin.

Schwarzmantel, J., 1994, *The State in Contemporary Society. An Introduction*, Hemel Hempstead: Harvester Wheatsheaf.

Seton-Watson, H., 1977, *Nations and States. An Enquiry into the Origins of States and the Politics of Nationalism*, London: Methuen.

Short, J.R., 1989, Yuppies, yuffies and the new urban order, *Transactions of the Institute of British Geographers NS* 14(3): 173–88.

Short, J.R., 1991, *Imagined Country. Society, Culture and Environment*, London: Routledge.

Short, J.R., 1993, *An Introduction to Political Geography*, 2nd edition, London: Routledge.

Short, J.R., 1996, *The Urban Order. An Introduction to Cities, Culture and Power*, Cambridge MA: Blackwell.

Shurmer-Smith, P. and K. Hannam, 1994, *Worlds of Desire, Realms of Power. A Cultural Geography*, London: Edward Arnold.

Sibley, D., 1995, *Geographies of Exclusion. Society and Difference in the West*, London: Routledge.

Slowe, P.M., 1990, *Geography and Political Power*, London: Routledge.

Smith, A.D., 1986, *The Ethnic Origins of Nations*, Oxford: Blackwell.

Smith, A.D., 1991, *National Identity*, London: Penguin.

Smith, A.D., 1995, *Nations and Nationalism in a Global Era*, Cambridge: Polity Press.